Logics of the Kingdom

A Scientific Analysis of the Word of God

Heather Fortinberry

ISBN: 978-1-7352587-2-0 (Print)

978-1-7352587-3-7 (E-book)

Printed in the USA

I dedicate this book
To all the wonderful young minds and hearts
With whom I've shared my knowledge as well as my love,
And to whom I've never lied.

Acknowledgments

Above all, I must give all credit to the *"three that bear record in heaven, the Father, the Word, and the Holy Ghost, and these three are one."* And to the *"three that bear witness in earth, the Spirit, and the water, and the blood: and these three agree in one"* (1 John 5:7-8 KJV). I must also thank all those who have dedicated years to studying and uncovering the mysteries of the universe, including but certainly not limited to: Plato; Galileo Galilei; Albert Einstein; Isaac Newton; Pierre Curie and his wife, Marie Sklodowska Curie; Antoine Henri Becquerel; Max Planck, Neils Bohr; Louis De Broglie; Richard Feynman; Steven Hawking; Paul Dirac; Ilya Prigogine; and Edward Lorenz, as well as all those who have been cited throughout this book. Without these individuals' hard work and dedication to their specific fields, I would have had nothing to which I could bridge the Word of God.

Contents

Preface

I thought as a child.
In the polite aftermath
Of idle chatter,
Adults would question,
"What do you do?"
"I think"— my simple reply.

Nothing much has changed. I wrote a poem in third grade, beginning with the words, "Sit and think all by yourself and imagine what to be." Somewhere it sits behind the same glass with red cloth tape around its edges as a testament to the fact that, truly, I haven't changed that much, at least with regards to the deep satisfaction that thinking has provided me. It's the tossing and turning of ideas and concepts in ways that make them more personally meaningful that has always generated inner delight. Words and ideas have been like candy wrapped in exotic foil, urging me tear them open as if with a promise of a unique surprise. And the last three years have been like visiting a candy factory for me, as I have been immersing myself in the knowledge bases of all the great Nobel Prize winners as well as those who have aspired to gain one themselves.

In the autumn of 2009 after realizing that I couldn't afford the doctorate program into which I had been accepted, I decided to begin researching for a new program of study that I might create, primarily with the idea of bridging psychology with the Word of God. I had been teaching various psychology courses for seven years at that point, and I had committed myself to living by the Word of God in 1985, two years after a car accident and a subsequent coma and brain damage (see *Paths of Resistance*). I felt that I had a decent enough handle on both disciplines to begin the process of systematically devising a program of study or at least a book, if nothing else. Without a doctorate in hand, my prospects for teaching appointments were limited, and this seemed to be a feasible way to provide me with some level of validity. So, I began to think and pray.

In life, everything seems to be connected; there is something called circular causality that establishes the interrelatedness of the myriad components that surface in any situation. Circular causality admits to the limitations imposed by evaluating a single component of a system without observing it in conjunction with the other components in the system. This thought led me to think about people's levels of understanding as well as the myriad influences that prevented people from acquiring greater understanding in all areas of life, but especially in the area of spiritual truths and the subject of God. I realized that one of the primary obstacles is that so many of the things pertaining to God made no logical sense, with an emphasis on the word, *logical*.

Accepting the notion of God requires faith, but that there was no scientific proof disturbed me. We have become a world that wants verifiable answers, and we want those answers to be reasonable; we want to be able to intellectually apprehend the things that we will support. As a teacher myself, I am not exempt from this desire; I want to understand things to the fullest possible level so that I can answer questions in a reasonable way that makes sense to my students, particularly. This is one of my own primary motivations in life, and my own need to understand made me look deeper into the fundamentals of faith as well as into the reasons why I, or anyone else, might believe in God.

The entire direction of my life changed in one afternoon's visit to Barnes and Noble. I had been stymied on a few occasions as I sought definitions of words that were not present in my own dictionary. I had a purpose when I entered the store; I wanted to obtain a new, "better, more fully equipped" dictionary. However, what I had envisioned wasn't there, so in disappointment I glanced down the adjacent isle. Suddenly my eyes landed on a section that was as foreign to me as Swahili-physics, and I felt that familiar pull that has presented itself whenever a certain path that I must follow is before me. Not knowing anything about physics, I picked up a friendly looking book and hurried home with the anticipation of finding treasure, and *that* I did.

That autumn day changed the course of my life; I began to see what I had always suspected—that everything in the physical world is actually substantiated in the Word of God and/or in the real-life experience of a true born-again believer. I have an unwavering faith in the truth of God's Word, so I knew that if He created the heavens and the earth, there must be some sort of evidence in science that would verify this. I knew that if I could just get a handle on what scientists, physicists, and chemists have understood, or are in the process of understanding, that a very logical bridge might be drawn between what God has said in His Word and what the world of science has come to believe.

As Jesus came to end the division between Jews and Gentiles, offering us both an opportunity to join His family through faith by the grace of our Lord Jesus Christ, so too do I offer an opportunity. The theory of *Logics of the Kingdom* (*LotK*) sheds light on the bridges between scientifically verified "facts" and the "truth" of God's Word, as written in the King James Bible, new and old. Four years of prayerful analysis and silence and an estimated 8,000 hours of studying has garnered enough evidence to at least place reasonable doubt on the perspective that what the Bible says is only a myth.

The atheistic existential idea that God has been created for men and by men for the simple purpose of providing men with meaning may in the end be logically questioned. In actuality, it would appear that everything that man has discovered through science and logic seems to line up perfectly with everything that the Bible proclaims had to have occurred in the very beginning, as well as what occurs in the fifth dimension of faith that is experientially verified in the lives of believers. Romans 1 aptly addresses this: *"What may be known of God is manifest in them, for God has shown it to [men]. For since the creation of the world [God's] invisible attributes are clearly seen, being understood by the things that are made, even His eternal power and Godhead, so that [unbelievers] are without excuse"* (Romans 1:19-20).

In essence, everything that is visible has been patterned according to pre-established principles that are already in existence, yet in an invisible, conceptual realm, as God told Moses, *"See that you make all things according to the pattern shown you on the mountain"* (Hebrews 8:5). Everything in this world can be thus discerned as adhering to a patterned substance or process that conceivably was created in the very beginning by God Himself. But don't simply take my word for it. Do your own research.

While the book in your hands was written for the scientifically-minded, I want to encourage you to open your heart and mind to gain insight into the patterns reflected in earthly processes that reflect those in heaven and have been revealed in scripture. As it says in Matthew 7:7-8, "*Ask, and it will be given to you; seek, and you will find; knock, and it will be opened to you. For everyone who asks receives, and he who seeks finds, and to him who knocks it will be opened.*" Believe that understanding is available by faith!

What I have encouraged my students to do is what I would encourage any reader of this work to do: Educate yourself and listen. Don't settle for information that comes from only one source. Critically analyze all points of view, but most importantly, be aware of your own heart's response. Do you have peace? Or do you fear men's judgment? Be honest with yourself and be willing to hear the messages that come to you in silence, when the world's drone is not drumming in your ear. Listen and watch. "*He who has ears to hear, let him hear!*" (Matthew 11:15).

Chapter 1
LotK Alpha

Without a doubt, the beginning has been one of the most debated subjects known to mankind. And it is debated for no other reason than the fact that the theoretical frameworks that have been developed, outside of what is spoken of in the Bible, have been plausible; their perspectives have been reasonable and are in line with other accepted theories in the scientific world. With regards to the beginning, what is not in question is the fact that our universe actually had one, contrary to the proposition that perhaps it is, has been, or always will be, eternal.

Interested in the notions of time and space, physicists Steven Hawking, George Ellis, and Roger Penrose applied Einstein's theory of relativity to these properties. Back in 1968 and 1970, the implications of Einstein's theory with regard to measurements of time and space led these men to publish papers concerning their findings. It turned out that according to their calculations, space and time had a definitive beginning that corresponded to the origin of energy and matter. Astronomers and physicists have since discovered tell-tale signs that there was indeed a point in time when nothing existed. In response to that, different ideas regarding the beginning of the universe have been formulated. However, when all is said and done, there is general agreement that a major event, like the big bang with its corresponding expansion, took place about 13.7 billion years ago.

As the prevailing theory to which most scientific theories seek to align, the big bang theory maintains that the universe was once very hot and dense, which caused it to expand. This expansion, in turn, caused the universe to progressively become cooler until it reached a temperature where the heat energy could be converted into protons, neutrons, and electrons, the constituent elements of all matter. The question of the original source of heat that caused the universe to expand remains unanswered, but the theory holds that the universe came into existence as a "singularity," or a point with infinite potentiality.

When this notion of singularity is considered, what may help our understanding is to consider it mathematically. In these terms, a singularity would be as a "point" with zero dimensions, which implies a space that the human mind cannot wrap itself around, similar to the notion of zero divided by a whole integer. According to our knowledge base, this simply doesn't work, but due to the inability to arrive at a quantitative answer, it remains in an open potential realm. And it is this conceptual potential that theorists consider relevant to the inception of the universe. Since the universe has been proven to have a beginning, the obvious creation of the wonders and magnificent beauty of the world, which has come into existence since that point, describes that point as having birthed infinite potentialities. It is with this reasoning, therefore, that the big bang theory proposes that the universe came into existence as a singularity.

What appears to be implicit is that this singularity, this point of infinite potential, had within itself the *seed* of heat and/or light, which eventually enabled the universe's expansion. Interestingly enough, however, while this proposed singularity harbored the seed of light, another example of a singularity is the core of a black hole, which would appear to harbor the seed of darkness. Instead of reflecting the properties of the objects that draw near, as does light, a black hole absorbs anything, including any frequency of electromagnetic radiation that approaches its outer ring or its "event horizon."

In essence, two separate and opposite examples of singularities emerge: One with electromagnetic-emitting properties and another with electromagnetic-absorbing properties. While both may aptly signify areas of unknown and perhaps infinite potential, one would aptly describe what the Bible proclaims as a domain belonging to "outer darkness" while the other would appear to illustrate what we know of God, a singularity having within itself the seed of light, behaving as does a fruit, like the fruit of the Spirit—love, as mentioned in Galatians 5:22, and as the descriptor of God: *"For God is love"* (1 John 4:8).

While the big bang theory maintains that the universe came into existence as a singularity, what does the Bible say about the beginning? *"In the beginning God created the heavens and the earth. The earth was without form, and void; and darkness was on the face of the deep. And the Spirit of God was hovering over the face of the waters. Then God said, 'Let there be light'; and there was light"* (Genesis 1:1-3). The Bible's explanation of the beginning obviously describes God as the singularity. And as a singularity that presumably had within itself the seed of light, what becomes clear through the Bible's presentation is the way in which that seed was disseminated—through God's words. God broke through the evident darkness and gave birth to light by speaking words that corresponded with His vision. He wanted light, and therefore He spoke, "*Let there be light,"* and there was light.

With reference to Hawking's, Ellis's, and Penrose's data that concluded that the universe had a definitive beginning, this point in time corresponds to the origin of energy and matter. Interestingly enough, energy and matter is contingent on the presence of light, and therefore light had to have presented itself first. In confirmation of Genesis 1:2, which asserted that "*darkness was upon the face of the deep,"* a definitive beginning that corresponded with the origin of energy and matter presumes that light had to have presented itself *suddenly*, and in stark opposition to, what was already present—in other words, darkness.

With regards to light's *sudden* appearance, this also appears to support God's way of performing His will. According to Malachi, *"'Behold, I send My messenger, and he [John the Baptist] will prepare the way before Me. And the Lord, whom you seek, will* ***suddenly*** *come to His temple, even the Messenger of the covenant, in whom you delight. Behold, He is coming,' says the Lord of hosts*" (Malachi 3:1). And, *"Watch therefore, for you do not know when the master of the house is coming—in the evening, at midnight, at the crowing of the rooster, or in the morning—lest, coming* ***suddenly****, he find you sleeping"* (Mark 13:35-36). There are many mysteries concerning the operations of God, but that He manifests Himself and His works *suddenly* is plainly understood, at least by those whose lives have been miraculously touched by Him.

While God's operatives often appear suddenly, such presentations are the actual fruition of yet another process—the speaking forth of an idea or an object of one's intention. Genesis 1:2-3 says, *"And the Spirit of God was hovering over the face of the waters. Then God said, 'Let there be light'; and there was light."* According to this Scripture, God spoke a command, and the words that He spoke yielded the physical substance delineated by His words. But how did God's words physically transform into actual substance?

That the power of God is beyond the human intellect is an understatement; however, His power actually operates according to a system about which people have extensive knowledge. The manifestation of God's power appears to be primarily electromagnetic in nature, with the entire electromagnetic spectrum, the domain of His operations. On one end of this spectrum, radio waves travel and these waves are interpreted as sound, and at another point on this continuum travel visible waves that are received by the human eye. While many frequencies that lie on the continuum of this electromagnetic spectrum have been discerned, there are likely frequencies that are humanly inconceivable. And it is with this potentially inconceivable estate that God is capable of governing all signs and wonders that are manifested through His servants or vessels here on the earth.

According to Emmy Noether's theorem, because matter, which is composed of charged particles, is created by light and is governed by conservation laws, any perceived charge actually signifies a symmetry debt of light.[1] As symmetry is based on conservation, and the annihilation of oppositely charged particles creates light, any registered charge would thus point to that which was missing in an equation that would register as light. The phenomenon of light would

1. John A. Gowan, "Section 9 - Symmetry: Noether's Theorem and Einstein's 'Interval,'" Modified May 2011, accessed January 28, 2014, http://www.johnagowan.org/sect9.html.

therefore be considered complete with infinite potential, having within itself the capability of all creation, while charged matter would signal an incompletion.

Interestingly enough, John 1:8 personifies the light; with reference to John the Baptist, the Scripture says: *"He was not that Light, but was sent to bear witness of that Light. That was the true Light which gives light to every man coming into the world"* (John 1:8-9). And further description: "*And the Word became flesh and dwelt among us, and we beheld His glory, the glory as of the only begotten of the Father, full of grace and truth. John bore witness of Him and cried out, saying, 'This was He of whom I said, "He who comes after me is preferred before me, for He was before me"'"* (John 1:14-15). This, of course, was referencing Jesus, whose alternate identity is the Word of God, who was in the beginning with God and was God (see John 1:1).

According to this Scripture, the Word of God appears to behave as does the supposed singularity referred to by the theorists. This singularity, however, was equal to God and according to John 1:4 had the life within Himself. This life, in turn, was equivalent to the light of men which shined in the darkness that had no understanding of it (see John 1:5). According to Genesis 1:3, this light came into being when God said, "*Let there be light*." What appears to have occurred is that after these words were spoken, light became the phenomenon that it is. We'll discuss this later, but with regards to the manifestation of light and everything else that God spoke into being in the beginning, one part of the equation may involve the coupling of phonons and photons, or the coupling of sound and light.

As light creates matter, and sound in the form of words creates a mental image of matter, the coupling of phonons and photons would appear significant. While phonons refer to the discrete modes of vibration within solid matter that enable our ears to hear sound, photons are the basic units of electromagnetic energy that provides us with light that we can see. We often consider these separately, but it has been found that under certain conditions, couplings of these two distinct phenomena may emerge. Referred to as polaritons, these couplings move through space with zero internal resistance to fluid flow. They also have superconducting properties, which means that they have zero *electrical* resistance. Without resistance to these two dynamics, polaritons thus experience an undiminished freedom and are not subject to being bound, or in other words, are not subject to being restrained from fulfilling their earthly potential. In terms of these couplings' significance, it may be interesting to note that they often result when an infrared photon and an optical phonon are mixed, exist in an ionic crystal or salt, and become excited by infrared radiation.

Generating heat rather than light, infrared radiation that exists in darkness may provide a domain wherein light energy becomes stored, which is then able to be retrieved when precise conditions are met. Nonetheless, that salt appeared prevalent is not without significance, particularly with reference to both a covenant of salt mentioned in Numbers, "*It is a covenant of salt for ever before the Lord unto thee and to thy seed with thee*" (Numbers 18:19 KJV), as well as to references to the disciples of Jesus Christ being the "*salt of the earth*" in Matthew 5:13. It would appear that when the patterned dynamics of salt appear on the scene, which implies the presence of a covenant relationship with God, both sensory barriers as well as resistant barriers become capable of being overcome.

Interestingly enough, in speaking about the beginning Genesis 1:2 says, "*And the Spirit of God was hovering over the face of the waters.*" While fresh water may have been present, the majority of water from the oceans and seas is salt water, with an abundance of sodium chloride flowing throughout. This would seem to indicate that both the fluid and conductive properties of salt water provided a space in which infinite physical as well as perceptual potentials were made possible. In essence then we may consider the notion that when light's creative ability is coupled with the vibrational waves of optical phonons in the medium of salt water, the stage is set for the realization of an infinite potential, and the heavens and the earth may very well have sprung from this platform.

The presence of sodium chloride in salt water would thus have provided a means for the necessary creative excitement to have occurred so many billions of years ago. And this excitement was not limited, as is implied by Genesis 1:2. Water was everywhere; and with the recent findings of water particles and/or ice on the Moon and on Mars, this claim appears to be substantiated. Water molecules on the Moon's surface were detected by three spacecraft—India's Chandrayaan-1 lunar orbiter, NASA's Saturn-bound Cassini probe, as well as its Deep Impact comet mission. The properties of water were also found to exist on, in, or around Mars, and in 2004 when NASA's Spirit and Opportunity rovers landed on Mars, they obtained evidence that Mars was both warmer and wetter billions of years ago than it is today. Vast reservoirs of water were found in the interior of Mars with some areas as wet as Earth's interior. In addition, the mantles of meteorites that had landed on Earth were analyzed by a technique called secondary ion mass spectrometry, and a determination was made that meteorites contained about the same percentage of water as Earth does.

On July 22, 2011, a cloud of water vapor containing 140 trillion times more water than the combination of all of the Earth's oceans was reported as having been discovered around 12 billion light years away from Earth.[2] Accordingly, it would appear that water has been prevalent in the universe for nearly its entire existence. While solar protons (the nuclei of hydrogen atoms) have been

2 Brian Dunbar, Whitney Clavin, and Alan Buis, "Astronomers Find Largest, Most Distant Reservoir of Water," NASA, July 22, 2011, accessed January 24, 2014, http://www.nasa.gov/topics/universe/features/universe20110722.html.

implicated in the physics of the production of this water, at least on the Moon, this theory relies on the Sun having been in existence. There appears to be no substantiated theory, however, as to how the Sun first came into existence. Genesis 1:16, however, does provide an answer: *"Then God made two great lights: the greater light to rule the day, and the lesser light to rule the night. He made the stars also."*

Men have for decades held the discovery of water on the Moon as a Holy Grail, believing that its discovery might signify the planet's potential for sustaining life, whether in the past or in the future. But the presence of water molecules on the Moon, as well as on Mars, signifies perhaps more than anything a need to question whether these findings are not truly remnants of the waters that are spoken about in Genesis. Now, of course, these are only two other planets, in addition to the cloud of water vapor, on which water has been firmly identified, but perhaps future space travels will find evidence of water on other planets or in other areas as well. And perhaps then, the waters of Genesis may be seen in a new light, providing a posthumous Nobel Prize to Thales (640-546 BC), a philosopher belonging to the Ionian League of ancient Greece. He held tightly to the notion that water was the first fundamental element of life, due to its intrinsic value.

In such a case that salt water was in fact everywhere in the beginning as the Bible proclaims, one thing is clear: There would be evidence of sodium chloride crystals having been present in abundance. Under the condition that the universe was also dark at that point, an opportunity would have appeared to avail itself: The coupling of optical phonons (internal vibrations) and infrared photons (photons that are generated from heat rather than light, whose hotter frequencies correspond with light's faster frequencies) known as polaritons. These couplings, in turn, could have conceivably provided the avenue through which thunderous vibrations, as well as lightning, traveled. Providing a substrate upon which invisible substance may be deposited, these couplings likewise exist in the invisible realm, and thus provide for us an example of a phenomenal platform that may have functioned as a major conductor of both the light and sound that appears to have occurred—in the beginning.

In addition, the appearance of lightning with the consequent coupling of thunder is alluded to in the Bible as pertaining to the voice of God. According to Psalms 77, the voice of God, and therefore His Words, thundered: *"The voice of Your thunder was in the whirlwind; the lightnings lit up the world; the earth trembled and shook"* (Psalms 77:18). While this may be considered simply allegorical, it's interesting to note that the second reference of thunder and lightning occurs in Exodus 20, after the Ten Commandments were issued forth: *"Now all the people witnessed the thunderings, the lightning flashes, the sound of the trumpet, and the mountain smoking; and when the people saw it, they trembled and stood afar off. Then they said to Moses, 'You speak with us, and we will hear; but let not God speak with us, lest we die.' And Moses said to the people, 'Do not fear; for God has come to test you, and that His fear may be before you, so that you may not sin'*" (Exodus 20:18-20).

As the Ten Commandments still appear as the basis of sound wisdom in terms of human behavior, the fact that thunder and lightning appeared at this time and was referenced as connected to God's voice might cause us pause. Along similar lines, the first reference was in Exodus 9:28 in response to Pharaoh having had enough of the hail storms that coincided with Pharaoh not letting Israel, God's chosen people, go free. What is apparent from these Scriptures is that thunder appears to be equal in measure and potency to words spoken presumably by God Himself; and words, like thunder, appear to be coupled with light. As words are emitted as sound waves, which are simply oscillations of pressure, what's interesting is the fact that thunder (sound) is simply a response to lightning's effect on the pressure and temperature of the atmosphere. Therefore, a correlational relationship appears to exist.

Using thunder and lightning as a foundational premise, what may not be recognized is that for all sounds there is likely a corresponding light wave that has preceded it. In consideration of this connection between sound and light, therefore, when something is heard, we may assume that a light frequency has already gone forth, albeit in the invisible realm. While sound registers in a frequency range to which humans are sensitive—between 20 Hz (cycles per second) to 20 kHz (20,000 Hz)—because of the slower speed of sound, the light to which it is connected will often be unrecognized by the human eye. As the wind which blows, yet remains invisible, has noticeable effects on the objects it passes, so too do the frequencies of the electromagnetic spectrum. While many of these frequencies remain beyond human detection, and outside of the realm of our awareness, they may have a substantial effect on our lives, leaving us unknowingly influenced by them.

As light is an electromagnetic energy wave which travels at a constant velocity of 186,282 miles per second, and the speed of sound is determined by the properties of the substance through which it travels, sound, obviously, is a fraction of the speed of light. In the case of thunder and lightning, thunder is essentially a response to lightning's effect; similarly, all sounds are likely to be simple perceptual responses to light emissions at different frequencies. The phenomenon of thunder and lightning, including the order in which they are perceived, has become apparent to us only because there is no larger natural power generated that registers to our human senses.

Conclusively, while the theoretical premise of the big bang appears to be valid, the question of the actual origin of the heat or light that caused the initial expansion of the proposed singularity still appears to be unanswered. And while no verification of the specific way in which the light/heat which caused the presumed expansion to occur is provided, perhaps as some have hypothesized, another part of the universe could account for the light's/heat's inception. Along with this proposed spatial entity, perhaps other missing masses will be located in another dimension or sphere. It's a possibility. However, with consideration of what the Bible has said, together with other scientific evidence, God's Word seems to illuminate much more on the subject. As the universe has been proven to have had a beginning with that beginning marked by the inception of light, the Word of God seems to accurately describe this period of time as well as allude to the power of acoustic vibrations, as it says in Genesis 1:1-3: *"In the beginning God created the heavens and the earth. The earth was without form, and void; and darkness was on the face of the deep. And the Spirit of God was hovering over the face of the waters. Then God said, 'Let there be light'; and there was light."*

Chapter 2
Water of the Word

In support of the Bible's proclamation that the *"Spirit of God moved upon the face of the water"* in Genesis 1:2, so too does *Logics of the Kingdom* propose the theory that the existence of water preceded all other physically verifiable substances. With the evidence that water, in its various phases, exists on or in the atmosphere of at least three celestial bodies—Earth, the Moon, and Mars, as well as in celestial space—this proposal appears to be substantiated. Another question, however, relates to water's ultimate significance. While meaninglessness and chance do enjoy a cultural spotlight, human nature, like the elements in nature, do tend toward order, which always presupposes a grander purpose for such alignment. As birds fly in an ordered formation for aerodynamic effectiveness, water's ubiquitous and foundational existence would appear to have a great significance for the purpose of carrying out some universal order. And with respect for universal order, we might first look up toward the Sun, the Moon, and the stars.

Beyond its notable ability to sustain life, the substance of water has also been credited for enabling the formation of stars. Comprised of hydrogen and helium gas as well as dust, stars are understood as emitting radiant energy that has been caused by sustained nuclear fusion. In order for stars to form, however, the implicit heat generated by the nuclear fusion required cooling; and this is where water's grand purpose would perhaps appear to reside—in its ability to cool atmospheric conditions by colliding with gas particles and absorbing these particles' energy, which would later be released in the form of radiation.

However, while there is evidence that stars were brought forth in a watery environment, the question of whether stars preceded water or water preceded stars comes into play. While stars apparently needed water's cooling properties in order to complete their stellar formation, according to National Geographic Creative, there is also evidence of a star 750 light years away that emits jets of water into space at 124,000 miles per hour. Water, therefore, also appears to be generated from stars, having apparently begun as simple hydrogen and oxygen atoms that formed on and around the star, becoming embedded in gas streams; and when they are far enough away from the star and rapidly cool, they end as jet streams of water, spewing forth from the star.[3]

Discovered by ESA's Herschel Space Observatory, this particular star leads us to believe that a vast quantity of water may actually be distributed throughout the universe from stars themselves. This not only poses the question of which came first, stars or water, but it also provides an illustration of a reasonable cyclical process, a self-perpetuating "seeding" process, which sheds light on the role that water may have played in the formation of the bodies within our own galaxy. If water were able to seed stellar formations, then the Bible's account in Genesis 1:2 of *"the Spirit of God hovering over the face of the waters"* would appear to be chronologically accurate. Accordingly, the Sun, the Moon, and the stars would have had a verifiable substance from which they may have spawned. In addition, with the evidence that a cyclical self-perpetuating seeding process may occur, it would appear that this serves as a pattern which is actually replicated on innumerable scales throughout the universe.

There is evidence, therefore, that while water appears to be a necessary component in the initial production stage of stars, it also appears to be a stellar product. As a product, water is said to be the result of the impactive interaction between an outflow of gas and dust that accompanies the birth of a star and the surrounding gas. As a result of the impact, abrupt changes in the pressure occur and produce shock waves. As disturbances that propagate through the material, these shock waves, in turn, compress and heat the gas, which ultimately creates water from the erupting mixture of hydrogen and oxygen.

According to this illustration, any potential significance of water as a product would only support the wonders of chance encounters and environmental conditions, which would simply echo the dominant cultural narrative: "There is no objective, universal meaning and purpose; all meaning and purpose is simply a product of human construction." According to these assumptions, water then appears only as a byproduct of gaseous interferences. There apparently is no purpose attached to its formation, other than the life-sustaining potential that it had that may or may not have come in handy at some point in time. That is, of course, if perchance the molecular climate achieved an appropriate state.

If we consider water as a product of stars and also as something that is perhaps required for their formation, a pattern appears to take shape that seems to mirror that which was described by Ilya Prigogine, as "dissipative structures." These "structures," in turn, then refer to the growth patterns that evolve and take on newness of life after an old structure has decayed or died. In this particular

3 Andrew Fazekas, "Star Found Shooting Water 'Bullets,'" National Geographic Daily News, June 13, 2011, accessed January 24, 2014, http://news.nationalgeographic.com/news/2011/06/110613-space-science-star-water-bullets-kristensen.

light, water may in fact appear as a structure that comes into being after a prior form has undergone decay; in this way, water may accurately symbolize a newness of life, a product of regeneration, renewal, and rebirth. As one who came to prepare the way of the Lord and make His paths straight (see Matthew 3:3), John the Baptist baptized with water unto repentance, which symbolized an initial renewal. Matthew 3:11 confirms this: *"I [John the Baptist] indeed baptize you with water unto repentance, but He who is coming after me is mightier than I, whose sandals I am not worthy to carry. He will baptize you with the Holy Spirit and fire."*

While regeneration, renewal, and rebirth appear to signify a spiritual awakening—as John contends, *"Most assuredly, I say to you, unless one is born again, he cannot see the kingdom of God*" (John 3:3)—the concept of decay and/or death before passage from one state into another, appears implicit not only in the spiritual realm but also in the physical. Herein lies the theoretical underpinning of the concept of gravitational collapse, through which everything from clusters of galaxies to planets and stars is considered to have come into existence. Upon such a concept lies the structure formation of the entire universe with an eventual ordering of molecular chaos into a hierarchy of structures. What is alluded to is a concept that adheres precisely to Ilya Prigogine's dissipative structures. Accordingly, order, or life, can make an appearance out of chaos, or death.

According to a meaningless paradigm or narrative, there would be no purpose attached to water's formation, other than the life-sustaining potential that may or may not have come in handy at some point in time. However, as a point of interest, let us just consider what the Word of God says about the potential worth, value, and purposeful meaning of the substance of water.

Aside from the oceans, seas, and rivers wherein water flows throughout the earth, water is also found to exist deep within the earth, which figuratively represents the deep things of God, as mentioned in 1 Corinthians 2:10: *"But God hath revealed them unto us by His Spirit: for the Spirit searcheth all things, yea, the deep things of God"* (KJV). As water washes over a substance and cleans those areas that were once caked with debris in order that each crevice is able to be fully examined, so too does the Holy Spirit reveal the hidden things of the heart. And it is this representative meaning of water that would appear to provide water with its ultimate purposeful value.

An example of water signifying the Spirit of God is evident also in John 4, wherein a Samaritan woman approached Jacob's well, ready to retrieve water. When this woman arrived at the well, however, Jesus was there, weary and tired from traveling. He asked her to retrieve some water for Him. Her response was of slight amazement as she questioned Him, *"How is it that You, being a Jew, ask a drink from me, a Samaritan woman?"* (John 4:9). As none of us are detached from our past experiences, psychologically speaking, she too responded accordingly. Due to social demographic prejudice, Jews typically reacted to Samaritans with disdain, not unlike many people's reaction to the homeless or the downtrodden, and she was therefore stunned that He would be asking *her* for water.

However, Jesus's response to her in John 4:10, unlike what she expected, clearly indicated that the substance of water represented the Holy Spirit of God: *"Jesus answered and said to her, 'If you knew the gift of God, and who it is who says to you, "Give Me a drink," you would have asked Him, and He would have given you living water.'"* And in John 4:13-14, Jesus answers: *"Whoever drinks of this water will thirst again. But whoever drinks of the water that I shall give him will never thirst. But the water that I shall give him will become in him a fountain of water springing up into everlasting life."* Among water's other attributes, therefore, it would appear that water's true significance is as a figurative representation of "*living water*" or the Holy Spirit that abides in the eternal and thus spiritual dimension.

While I speak of water's ultimate significance or purpose, let it be understood that there is but one purpose in all of life that has any significance—to glorify God. Everything honorable or dishonorable has but this one purpose. That having been said, it would seem that water's ultimate significant purpose was to provide a means by which God may be glorified, and with the glorification of God being intricately tied to the fulfillment of His Word, water's significance would appear to be tied to the impartation of the Holy Spirit in the lives of believers.

With water as a representation of the Holy Spirit, we may see the fulfillment of Jesus's death and resurrection as a primary way by which both the Word of God was fulfilled and by which God was glorified. Without Jesus dying on the cross and being resurrected the third day, God's Word would not have been fulfilled, which said, "*Behold, we go up to Jerusalem; and the Son of man shall be betrayed unto the chief priests and unto the scribes, and they shall condemn him to death, and shall deliver him to the Gentiles to mock, and to scourge, and to crucify him: and the third day he shall rise again*" (Matthew 20:18-19 KJV; see also Mark 8:31; Luke 9:22).

With the promise of rising again, Jesus was obedient unto death and thus enabled the Holy Spirit to come live within the hearts of those who believe on Jesus' name: *"And I will pray the Father, and He will give you another Helper, that He may abide with you forever—the Spirit of truth, whom the world cannot receive, because it neither sees Him nor knows Him; but you know Him, for He dwells with you and will be in you. I will not leave you orphans; I will come to you. A little while longer and the world will see Me no more, but you will see Me. Because I live, you will live also"* (John 14:16-19). This Scripture provides the referential evidence of the Spirit as that which would follow Jesus and live *in* believers, providing a source of "living water."

Answering the question of what makes this water "*living,*" it becomes essential to consider the substance after which this is patterned. While the Holy Spirit is one manifestation of God, with the Father and the Son/Word as the other two, there is a natural process that operates similarly here on Earth. As a record bearer in heaven and as a witness on the earth (see 1 John 5:7-8), the Holy Spirit is the one aspect of God that appears to bridge both earthly and heavenly spheres. And as such, He would appear to behave as a transmitter of a broad spectrum of frequencies. And when we speak of frequencies, we are specifically referring to the domain wherein electrical signals travel.

As everything earthly is patterned according to a spiritual or heavenly substance, or process, this domain would appear to find equivalent expression in the substances known as ionic liquids, such as molten salts, which are fluids that conduct electricity, or electrolytes. Electrolytes are thus patterned after "living water," because they, like the Holy Spirit, are capable of transmitting electrical signals throughout a body. Accordingly, we can understand that when the Holy Spirit lives in those who have believed that Jesus is the Son of God and have accepted Him as their personal Lord and Savior, they are capable of transmitting the Holy Spirit's signals. Living within "believers," the Holy Spirit then enables them to connect with God directly, to hear His voice, and to receive spiritual gifts, as God is electrical in nature, with electromagnetism the domain of His power.

While the Holy Spirit is considered, in one sense, representative of the "living water" that Jesus spoke about, the Word of God is also an aspect of this water as referenced in John 4:13-14. This truth comes from the reasoning that there is total correspondence between all aspects of God. First John proclaims, "*There are three that bear witness in heaven: the Father, the Word, and the Holy Spirit; and these three are one. And there are three that bear witness on earth: the Spirit, the water, and the blood; and these three agree as one*" (1 John 5:7-8). In addition, the Holy Spirit is referred to as the "*Spirit of truth*" in John 14:17 and the Word is referenced as truth in John 17:17, "*Sanctify them by Your truth. Your word is truth.*"

Using deductive Aristotelian logic, therefore, both the Word of God as well as the Holy Spirit may be discerned to be the same. Both are manifestations of, and conduits to, the truth of God. And when the "ionic fluid" of the Holy Spirit exists within a believer, it will resonate with the Word of God and all truth with which it corresponds or agrees. Thereby we ourselves become witnesses to the truth. As a living entity, this "living water" or "Spirit of truth" is ultimately responsible for our connection to God the Father and His Son, Jesus Christ. As a record bearer in heaven and as a witness on earth (see 1 John 5:7-8), the Holy Spirit enables us to resonate with the truth of God's Word, which bears record that Jesus Christ is the Son of God. According to what is written in Romans 1:3-4, *"Jesus Christ our Lord, who was born of the seed of David according to the flesh, and declared to be the Son of God with power according to the Spirit of holiness, by the resurrection from the dead,"* and 1 John 5:10, *"He who believes in the Son of God has the witness in himself,"* the witness of the Holy Spirit provides a living testament to the truth.

While water is used as a figurative representation of the Holy Spirit, an interesting correlation exists between water, the Holy Spirit, and the property of life itself. With reference to the beginning, Genesis 2:5-7 says, *"For the Lord God had not caused it to rain on the earth, and there was no man to till the ground; but a mist went up from the earth and watered the whole face of the ground. And the Lord God formed man of the dust of the ground, and breathed into his nostrils the breath of life; and man became a living being."* The presence of water, therefore, became pivotal in providing life and a "living" existence for God's image-bearer, man.

Accordingly, the Holy Spirit, which Jesus' death and resurrection enabled man to receive, provides life and a true "living" existence, without which man is considered dead:

"*And you He made alive, who were dead in trespasses and sins, in which you once walked according to the course of this world, according to the prince of the power of the air, the spirit who now works in the sons of disobedience, among whom also we all once conducted ourselves in the lusts of our flesh, fulfilling the desires of the flesh and of the mind, and were by nature children of wrath, just as the others. But God, who is rich in mercy, because of His great love with which He loved us, even when we were dead in trespasses, made us alive together with Christ (by grace you have been saved), and raised us up together, and made us sit together in the heavenly places in Christ Jesus*" (Ephesians 2:1-6).

Therefore, beyond the cooling properties of water, the true significance and purpose of water would appear to illustrate the vital role that the Spirit, like water, has in providing man with life, which enables God to receive glory as a merciful and loving Father.

Chapter 3
Three as One

"For there are three that bear witness in heaven:
The Father, the Word, and the Holy Spirit; and these three are one.
And there are three that bear witness on earth:
The Spirit, the water, and the blood; and these three agree as one."
(1 John 5:7-8)

The three manifestations of God mentioned in 1 John 5:7-8 above all must be understood as having total agreement with one another, which would imply harmonious reflection. At no point in space or time would one's *essential* characteristics differ from another's. The word *essential* is stressed here due to the fact that non-essential characteristics do in fact differ, such as function and form, but the integrity of the core value remains intact. Therefore, while the Father, the Word and the Holy Spirit each have their own unique substantial form, functioning according to each unique office, their oneness can be understood in relation to each one's role in fulfilling the preordained, written Word of God, and in glorifying the God of Abraham, Isaac, and Jacob and His Son Jesus Christ, the Messiah.

This, after all, is the grand purpose of why Jesus was sent, as Isaiah 61 confirms: "*The Spirit of the Lord God is upon Me, because the Lord has anointed Me to preach good tidings to the poor; He has sent Me to heal the brokenhearted, to proclaim liberty to the captives, and the opening of the prison to those who are bound; to proclaim the acceptable year of the Lord, and the day of vengeance of our God; to comfort all who mourn, to console those who mourn in Zion, to give them beauty for ashes, the oil of joy for mourning, the garment of praise for the spirit of heaviness*" (Isaiah 61:1-3a). Now, if we were to ask the question why, we would reasonably obtain the answer as to why Jesus, the Word, was promulgated to all those on earth. Why? "*That they may be called trees of righteousness, the planting of the Lord, that He may be glorified*" (Isaiah 61:3b). Jesus, upon whom the Spirit of God rested, was thus sent by God to give those who believe an opportunity to become as full grown, mature trees that bear the fruit of righteousness, with one primary intention—that God might be glorified.

Through the avenues of preaching and proclaiming, Jesus, the Word of God, opened the door to the spirit of wisdom and revelation that, if received, provided an opportunity for men's eyes of understanding to be enlightened, that they might know the hope of His calling (see Ephesians 1:18). These were the avenues by which faith came, as Romans 10:17 declares: "*So then faith comes by hearing, and hearing by the Word of God.*" And when wisdom, knowledge, and understanding are brought forth in a person's life, they can become as full grown, mature trees that bear the fruit of righteousness, and hereby these people are enabled to fulfill their God-given purpose and potential.

As trees that have been planted in good soil and have received the Sun's light bear the fruit that they were preordained to bear, so too do those who "see" and "hear" have the opportunity to bear the fruit of the Spirit that the children of God have been preordained to bear. Galatians 5:22-23 mentions these seed-bearing fruits that according to their nature produce after their kind: "*But the fruit of the Spirit is love, joy, peace, longsuffering, kindness, goodness, faithfulness, gentleness, self-control.*" When these particular fruits are born by people who have been planted by the Lord, they are planted for the purpose of glorifying the Lord, in proclamation of His perfect design and order. If the expected fruit is brought forth at the expected time, God is glorified because the tree's function operates according to the preordained vision.

In the example of the parable of the fig tree in Luke 13, when the tree that had been designed to bear fruit continuously bore no fruit, it testified of its own deceptive identity; it was as a hypocrite, outwardly showing itself to be one thing while functioning as another; looking alive with purpose, but behaving as one that was dead. God receives no glory in such a hypocritical situation, but is glorified instead when what He has designed produces its predestined fruit, when the end result aligns with the preordained vision. However as Habakkuk 2:2-3 implies when it says, "*Write the vision and make it plain on tablets, that he may run who reads it. For the vision is yet for an appointed time; but at the end it will speak, and it will not lie. Though it tarries, wait for it; because it will surely come, it will not tarry,*" one fruit is required above all—faith. For "*without faith it is impossible to please him: for he that cometh to God must believe that he is, and that he is a rewarder of them that diligently seek him*" (Hebrews 11:6 KJV). Faith requires persistent belief in a preordained vision in the face of contrary evidence, and faith is the means by which God is both pleased as well as glorified.

Accordingly, as the unifying purpose, or oneness, that binds the three aspects of God together, the glorification of God, the Father of our Lord Jesus Christ, requires faith. God is glorified when those who believe exhibit faith in God's Word and in the name of His Son, Jesus Christ. One of the original men of faith, Abraham, by whom we are all enabled to be members of the family of God, glorified God by believing the words of His promise. "*And being not weak in faith, he considered not his own body now dead, when he was about an hundred years old, neither yet the deadness of Sarah's womb: he staggered not at the promise of God through unbelief; but was strong in faith, giving glory to God"* (Romans 4:19-20 KJV).

By having faith in God's Word, Abraham glorified God, and by fulfilling His promise to Abraham, God fulfilled His own prophetic words to Abraham: "*This shall not be thine heir; but he that shall come forth out of thine own bowels shall be thine heir. And he brought him forth abroad, and said, Look now toward heaven, and tell the stars, if thou be able to number them: and he said unto him, So shall thy seed be. And he believed in the Lord; and he counted it to him for righteousness. And he said unto him, I am the Lord that brought thee out of Ur of the Chaldees, to give thee this land to inherit it"* (Genesis 15:4-7 KJV). And herein lies one unifying record or witness—Father, Son/Word, and Holy Spirit will agree as to the legitimacy of the true land of Israel's possession, and the fulfillment of such a directive, as all other prophetic fulfillments, will both please and glorify God.

While glorification of God may be understood as a unifying purpose of His three aspects, a mystery remains as to why three manifestations of one God would even exist. To what end might this apparent division serve? After what kind of pattern may we understand this mystery? As everything in this world appears to have been made after a pattern in the heavens, with the Lord's Prayer in Matthew 6:10 and the building of the tabernacle in Exodus 25:40 serving as examples, the pattern of the triune God must find an earthly expression.

As the words of Lord's Prayer, "*Thy kingdom come; thy will be done in earth, as it is in heaven,"* imply, there is an opportunity for the earth and earthly activities to mirror heaven's. And according to God's admonishing, the pattern of the tabernacle was already complete before Moses received it in a vision; it quite clearly existed in the heavenly realm before that time. Decisively then, "*there is no new thing under the sun*" (Ecclesiastes 1:9 KJV), at least with regard to the fundamental pattern of things. The images of the patterns as well as processes of everything in, on, around, and under the earth, therefore, exist in a preordained state of completion in heavenly places. In like manner, the three parts of God serve as a pattern that will be evident in earthly processes or constructions.

Inasmuch as change has been identified as the only consistent thing in life, the subject of phase transitions becomes even more central to the topic of universal truths. If we reduce our scope to the micro level, we will verify that everything in nature undergoes phase transitions, where one state of matter transitions into another state, all the while maintaining its fundamental identity. While God is one and yet has three distinct aspects to His being, we may be able to understand these three aspects as three different phases of God. Using the example of a first order phase transition whereby solids, liquids, and gases transition into each other's phase state, the parameters of temperature and pressure determine the final state of density.

Occurring in a substance that can undergo magnetization, however, a second order phase transition is determined by an invisible, spontaneous parameter that occurs in the twinkling of an eye (1 Corinthians 15:52). In determining the difference between these two transitions, what becomes evident is that with first order transitions the transitional *process* may be "measured"; the substance could be released from the temperature conditions and the mixed states would become apparent. With second order transitions, however, the exact point of magnetization is beyond finding out, as it occurs in an instant, faster than the human eye can perceive. However, despite occurring beyond human perception, an immediate transition within the material is evident as it draws toward itself anything with magnetic potential. At no point can we actually *see* the point at which magnetism has occurred, but an immediate transition within the material is evident. The three manifestations of God, therefore, may be readily understood according to this concept of phase transitions.

John 1:14 speaks about a transition that occurred to the Word itself: *"And the Word became flesh and dwelt among us, and we beheld His glory, the glory as of the only begotten of the Father, full of grace and truth."* Many a skeptic has held this notion in contempt; after all, how is such a transition possible? Because everything in this world is made up of atoms, any spiritual phase transitional pattern will mirror what occurs in the physical realm. Accordingly, if we are to consider the Word, in substantial form, as analogous to water, then the process that would parallel the heavenly pattern would likely be freezing. As an exothermic process that actually releases heat energy, freezing produces a denser form of the substance under consideration, except in the case of water, which we know is denser in its liquid state. However, beyond this exception, exothermic processes which let go of, or loosen, heat energy from a substance will become denser.

In the case that the Word of God were to be considered analogous to spirit or vapor however, the phase transitional mode would be likened to deposition, where a vaporous form is deposited on a substrate, and initiates a new growth process. With regards to John 1:1, this consideration would lead us to conclude that the Spirit of God, like a vapor, was deposited on the substrate of a gamete in Mary's womb. Under both conditions, the process of the Word becoming flesh appears to coincide with a phase transitional mode where the substance of the Word gives up its inner heat energy. This release, then, of that which has provided it with a more ethereal form, begins the process of the substance becoming denser and more bound, which seems to speak perfectly of the human experience.

While deposition appears to be alluded to John 1:1, its opposite process of sublimation also appears in Scripture. Marked as a phase whereby a solid transforms into a vapor, bypassing a liquid phase, sublimation appears to be a process that coincides with the "purpose" of becoming less dense, less bound, and thereby more free. As an endothermic process which entails the absorption of heat energy, sublimation is a process whereby matter becomes less and less dense to the point of escaping its solid confines. As deposition appears to be the means by which the Word became flesh, sublimation appears to be the process by which, "*Enoch was taken away so that he did not see death, 'and was not found, because God had taken him'*" (Hebrews 11:5), as well as is likely the process that will be experienced by those who, "*shall not taste of death till they see the Son of man coming in His kingdom*" (Matthew 16:28). Interestingly enough, the first order phase transitions that mirror these scripturally referenced processes are not readily discernible, and might rather appear as second order transitions that occur in the blink of an eye.

It would appear that with this understanding of how phase transitions relate to these foundational doctrinal truths, we may more readily discern the sacrifice that Jesus made for us in taking the form of mankind; He ultimately humbled Himself by making a willful transition into a phase of matter that was bound to the earthly elements that had been deposited into the Devil's domain. He gave up ultimate freedom and became as one bound, but He did it all for a determined purpose—that God might be glorified. He willfully and mindfully gave Himself as a ransom in exchange for the souls who might accept this gift of salvation. And let us not forget that Jesus was a man just like we are, who was at that time bound in a body with a soul and with a mind. Philippians 2:5-7 provides an account: *"Let this mind be in you which was also in Christ Jesus, who, being in the form of God, did not consider it robbery to be equal with God, but made Himself of no reputation, taking the form of a bondservant, and coming in the likeness of men."* Jesus was one with God, but *chose* to transition into another form—for us—with the ultimate intent of providing us with an opportunity to glorify God.

Note that aside from alluding to phase transitions, this Scripture accurately credits the mind as the governor of a man's purpose; for without mindful or deliberate thought, a body will simply follow every fleeting emotion or passion. A man's purpose, therefore, is determined by the stream of thoughts that have been allowed to gain momentum, for good or for bad. Jesus, therefore, purposed in His mind to fulfill God's will and to do all for His Father God's glory, which was to ultimately follow a preordained path that had been prepared for Him to follow.

This path upon which Jesus dared not stray would ensure that both Jews and Gentiles would have access to both the kingdom of God, which is the Holy Spirit, as well as to the kingdom of heaven, which is the operational mode by which God's will is accomplished. As the kingdom of God is within us (see Luke 17:21), the kingdom of heaven describes the government of God, or in other words, *His ways*. And as His ways are higher than our ways (see Isaiah 55:9), this government is revealed through parables, which requires discernment that comes most readily from the kingdom of God (Holy Spirit) living within. We may conclude therefore that the indwelling of the Holy Spirit, which agrees with Jesus/the Word, enables our minds to stay on the preordained path that has been prepared specifically for us. And if our minds remain on this preordained path and we fulfill the purpose proposed by the Holy Spirit, then God will be glorified, His will shall be done, and at the appointed time, we will potentially transition into another phase of matter as did Enoch.

Chapter 4
The Triple Point of God

While first order phase transitions would appear to follow the pattern of the phase transitions to which the Bible alludes, an element's triple point also stands out as serving as the basis of scriptural reference. As we have noted, when an element freezes and transitions into a solid phase, that phase is denser than the element's liquid state except in the case of water, i.e., ice floats in water. Now, in the case of water that covers more than half the earth and is the home of myriad species, we can cite purpose as a reason: Water's frozen blanket serves as a floating protective layer for all those living within the oceans and seas. This purpose serves as a reasonable answer to why water's giving up of heat energy produces a less dense phase.

Under the condition that water might serve as a representative figure of the Word of God, the Word of God would also become less dense should it likewise give up heat energy. How would the Word give up heat energy? It would be spoken; and if it were spoken, its intrinsic form would take on a protective quality as does a protective layer of ice, and that which released the Word would become less dense, less bound, and more buoyant.

From everything spoken thus far, we see that liquid water and ice are two phases that hold some level of mystery, but there's another phase of water that trumps these two—snow. Snow is special due to its intermediate state, somewhere between water and ice, but closer to frost; and until special processes were developed, the structural consistency of snow was not readily reproducible. An interesting Scripture comes to mind regarding snow in Job 38:22: *"Hast thou entered into the treasures of the snow? or hast thou seen the treasures of the hail which I have reserved against the time of trouble, against the day of battle and war?"* (KJV). Aside from the suggestion that God uses snow and hail as tools through which He may send a message—like thunder and lightning, i.e., *"Fire and hail, snow and clouds; stormy wind, fulfilling His Word"* (Psalms 148:8)—to what treasure may the Scripture allude? Intuitively, treasure is something that is valuable and worth searching for. The true value of snow, however, may in fact simply be the temperature at which it occurs; snow's particular structure often occurs at water's triple point.

Most of the elements have a triple point, which marks the temperature at which all three phases—vapor, liquid, and solid—may simultaneously coexist. Snow, however, occurring at water's triple point, appears in an entirely different structural form, and is a glorious representation of beauty with each flake miraculously symmetrical and unique, as are the people for whom Jesus died. This triple point of water then appears to be patterned after the triple point of God, a phase state that is apparent only at a particular temperature at which all three phases of God—the Father, the Word/Jesus, and the Holy Spirit—can simultaneously coexist. Accordingly, we may assume that this triple point delineates something to which all three aspects of God agree, for where there is agreement, there is coexistence. And with the consideration that the fulfillment of God's will, which is revealed in His testaments, provides Him with glory, the triple point of God may be rightly considered His Word and/or the bearers thereof.

"For as the rain comes down, and the snow from heaven, and do not return there, but water the earth, and make it bring forth and bud, that it may give seed to the sower and bread to the eater, so shall My Word be that goes forth from My mouth; it shall not return to Me void, but it shall accomplish what I please, and it shall prosper in the thing for which I sent it" (Isaiah 55:10-11). As water's triple point manifestation, snow is a molecular mirror image of all phases of H_20. While the temperature at which it occurs is readily identifiable, it signifies a phase to which all phases of water must agree, with its primary function being to be absorbed by seeds, so that growth might be achieved. Similarly, God's triple point phase would appear to be the Word of God, as everything pertaining to God must be corroborated by the Word, which has as its primary goal to be absorbed into the heart of mankind.

However, while the Word itself may be considered to be the primary triple point phase state, so too may the senders or apostles of the Word. Baptized in the Holy Spirit of God and armed with "*the sword of the Spirit, which is the word of God*" (Ephesians 6:17), those prepared with the "*whole armor of God*" (Ephesians 6:13) may be considered by God as treasure like the snow. Psalms 135:4 specifically refers to Israel in this context: "*For the Lord has chosen Jacob for Himself, Israel for His special treasure.*" And according to Romans 2, all those who "*do by nature the things contained in the law, these, having not the law, are a law unto themselves: which shew the work of the law written in their hearts, their conscience also bearing witness*" of the truth (Romans 2:14-15 KJV). Gentiles may thus be considered as Israel: "*For he is not a Jew, which is one outwardly; neither is that circumcision, which is outward in the flesh: but he is a Jew, which is one inwardly; and circumcision is that of the heart, in the spirit, and not in the letter; whose praise is not of men, but of God*" (Romans

2:28-29 KJV). Both Jews and Gentiles, therefore, may be considered as snow, the triple point of water, and treasure unto God.

While representative of treasure, snow is also mentioned in the context of having been reserved for the time of trouble: *"Have you entered the treasury of snow, or have you seen the treasury of hail, which I have reserved for the time of trouble, for the day of battle and war?"* (Job 38:22-23). And with regard to the Word of God assuming a triple point phase, 2 Thessalonians alludes: *"For the mystery of iniquity doth already work: only he who now letteth will let, until he be taken out of the way. And then shall that Wicked be revealed, whom the Lord shall consume with the spirit of his mouth, and shall destroy with the brightness of his coming"* (2 Thessalonians 2:7-8 KJV). The Word of God, which is Jesus, like the reserved treasure and "battle ax" of snow and its counterpart, hail, serves the function of appearing at a time or "temperature" when the fullness of its essential power needs to be made known.

Aside from any comparison between snow and the Word of God, however, what becomes evident from this last Scripture is that the Lord will both judge as well as consume the wicked with that which comes forth out of His mouth, or in other words, with His Word. And as Isaiah 45:23-24 declares: *"I have sworn by Myself; the word has gone out of My mouth in righteousness, and shall not return, that to Me every knee shall bow, every tongue shall take an oath. He shall say, 'Surely in the Lord I have righteousness and strength. To Him men shall come, and all shall be ashamed who are incensed against Him.'"*

As it appears that the three parts of God have in common a triple point phase state—the Word—this proposed triple point is analogous to a point of highest potential. Like snow manifests at water's triple point and has the potential of transitioning into any of the other phases, the Word of God, as a record bearer (1 John 5:7), holds the legal right to the securities which are possessed by all three phases of God, such as confidence, freedom from doubt, anxiety, or fear. As the parable of the leaven spoken of in Matthew 13:33 describes, a key operative of the kingdom of heaven appears to be a hidden unifying principle that brings all three *"measures of meal,"* or aspects of God, to a phase state where all three harmoniously coexist, and can at that point fulfill the loaf's true potential—to make all three as one.

Chapter 5
Facts versus Truth

The structural pattern of all that has been made on earth does indeed appear to emulate a spiritual prototype that exists in the heavenly realm, and hereby we may verify an intention spoken about in the Lord's Prayer. *"Our Father in heaven, hallowed be Your name. Your kingdom come. Your will be done on earth as it is in heaven"* (Matthew 6:9-10). One of the Father God's apparent intentions was to have the substances and processes on earth mirror the patterns established in the heavens. Accordingly, one of the primary reasons why the Ten Commandments were given to Moses becomes evident: These commandments illustrated the foundational truths upon which the government of the kingdom of heaven was established.

In providing the Ten Commandments to Moses and stressing Israel's obedient allegiance to them, God was sharing with Israel how they might live in the kingdom of heaven on earth, and how they might reap the full fruit of the covenant blessing that had been bestowed upon them. The covenant had been given to Abraham and to his seed after him, as is confirmed in Genesis 22:17-18: "That in *blessing I will bless you, and multiplying I will multiply your descendants as the stars of the heaven and as the sand which is on the seashore; and your descendants shall possess the gate of their enemies. In your seed all the nations of the earth shall be blessed, because you have obeyed My voice."* This call to obedience was once again brought forth as a message from God, by Moses before he had received from God the Ten Commandments: *"Now therefore, if ye will obey my voice indeed, and keep my covenant, then ye shall be a peculiar treasure unto me above all people: for all the earth is mine: and ye shall be unto me a kingdom of priests, and an holy nation"* (Exodus 19:5-6 KJV).

This Scripture, in addition to establishing obedience to the voice (the Word) of God as a condition of the established covenant, interestingly enough also supports the idea that God's children, as bearers of the Word, are esteemed as treasure; as each flake of snow appears perfect and unique, having traveled its own ordained path, so too will the children of God, who remain obedient to the Word of God. If they keep the commandments and water the ground of their hearts with the Word, they will exist in a state where their needs are met as priests, who while not given land for an inheritance are instead provided with the fruits of others' increase. While this particular operational mode defined the reality of the priests in the Old Testament, we will likely see this play out in the fulfillment of Proverbs 13:22: "*A good man leaves an inheritance to his children's children, but the wealth of the sinner is stored up for the righteous.*"

As this Scripture described the reality of Israel's priests of old, so shall it describe a transfer of wealth included in the blessing of God; and interestingly, as there is always a counterfeit or a polar force to the true, there is likewise now a push to steal wealth from the rich and give it to the poor under the guise of a just distribution. This, however, is a false reflection of the true transfer of wealth, which will invariably glorify God; and hereby we may discern the truth from mere facts. The truth glorifies God. However, if God and His Word are not glorified, what is heard can be regarded as a lie.

In response to this, you may say, "There are many facts that are true that do not necessarily glorify God; so this assertion is false." But I say, there is a definitive difference between fact and truth that makes this assertion true. What is true? Philippians 4:8 speaks about things that are true, and that along with things that are honest, just, pure, lovely, and of a good report, our thoughts should be trained to consider only these things as valuable. What determines whether something is a truth or a fact, therefore, is whether the evidence aligns with the preordained will and Word of God. What is truth? This was a question that Pilate had asked Jesus after Jesus answered his question, "*Art thou a king then?*" with, "*Thou sayest that I am a king. To this end was I born, and for this cause came I into the world, that I should bear witness unto the truth. Every one that is of the truth heareth my voice*" (John 18:37 KJV). What is truth? "*The voice of his word*" (Psalms 103:20 KJV).

This is a prime example of truth, because it bears witness to the prophetic words spoken in the Bible about Jesus, as John declares: "*And the Word became flesh and dwelt among us, and we beheld His glory, the glory as of the only begotten of the Father, full of grace and truth*" (John 1:14). "*For God so loved the world that He gave His only begotten Son, that whoever believes in Him should not perish but have everlasting life*" (John 3:16). The truth, therefore, is found in the Word of God, "*Your word is truth*" (John 17:17). It is also found in the Spirit of God: "*the Spirit is truth,*" and, "*the truth is in Jesus*" (1 John 5:6; Ephesians 4:21).

Nonetheless, facts that have been verified by man's eyes or by instrumentation devised by man's hands have slowly but surely eroded the structural strength and integrity of anything purported to be "true." All is now measured relatively, and the truth is, at best, treated as an attractive nuisance into which people fear falling without a well-devised plan of how to escape. The notion of truth as a concept has been slowly but surely wearing away like a wooden jetty, with its pylons eaten to their very core. However, we must remember that truth's foundational strength has been challenged since the initial temptation in the garden of Eden. Satan successfully used the *fact* that the tree of the knowledge of good and evil bloomed with appetizing fruit to seduce Eve and Adam into disobeying God's words. This *fact*, however, only distorted the *truth* of why God had commanded them not to eat of it. He wanted to get the point across early on that "*Man shall not live by bread alone, but by every word that proceeds from the mouth of God*" (Matthew 4:4). Accordingly, we may see the role that facts have in distorting the reception of the truth.

In the first place though, what compels us to believe facts? It's obvious; they have been substantiated, or given form, by empirical evidence. When facts are cited, the implication is that they can be verified by some sort of tangible evidence. However, the supposed substantial weight of this evidence is contingent on the accuracy of the measuring tool as well as the accuracy of the measurer. Nonetheless, even if both the measuring tool as well as the one who measures were without flaw, the evidence may in fact be understood to be far less valuable than ever before believed.

Consider this: If we were to apply the knowledge of the phases of matter that undergo phase transitional processes to notions such as facts and truth, facts would be represented as solids, and truth would be represented as vapor. Substantiated facts are known by the senses; they can be seen, touched, and understood, or stood under and thoroughly examined. Truth, on the other hand, cannot. For example, if one stood under an ocean, which represents a truth beyond our human grasp, with hopes of examining it, they would be fully overwhelmed. Accordingly, as solids are known by the senses, being bound in discernible matter, vapors are discerned only when an illuminating light is shed upon the atmosphere in which they freely flow. And while most solids are subject to time and thus decay, vapors appear to be free of physical corruption. According to this distinction between solid and vapor therefore, we may gain insight into the difference between facts and truth. Facts are bound by temporal knowledge; the truth, however, often requiring intuitive discernment, is free from and not bound to man's reasoning.

As with a vapor that escapes any solid confine, truth escapes the confines of any solid phase and becomes subject to the sphere of eternal life. And when we consider the phase transitional processes that either release or absorb heat energy, with the end being either freedom or a state of bondage, would it not be reasonable to conclude that a phase such as sublimation that absorbs heat energy—making things less dense, and potentially achieving the ability to escape its solid confines—could be esteemed of greater intrinsic value? With entropy and decay basically describing an exothermic process whereby heat energy is lost instead of gained, wouldn't processes that follow this pattern be considered less desirable?

According to this framework then, it would become apparent that "hard" or solid evidence is actually on a lower level with less intrinsic value than "soft" or vaporous evidence. With the flesh of men serving as an example of solid evidence and the Spirit of God within a believer serving as vaporous evidence, Romans 8 supports this reasoning: "*For they that are after the flesh do mind the things of the flesh; but they that are after the Spirit the things of the Spirit. For to be carnally minded is death; but to be spiritually minded is life and peace*" (Romans 8:5-6 KJV). When our thoughts are governed by the solid factual evidence of the fleshly world, therefore, the path drawn is toward death and not life. When our thoughts, however, yield to the Holy Spirit and to God's Word, the path that is drawn follows peace and leads to life everlasting. When we absorb the Word, which is truth, we become lighter and less bound to the flesh; and when we let go of the Word, as during a process of entropy, we become more bound to the flesh, and thus weighted down by material that intrinsically decays.

We live our lives according to many *facts* or pieces of information presented as having objective reality. However, it would be beneficial to consider that scientifically based discoveries are continuously transforming our understanding of what is real. Our paradigms shift all the time in relation to new knowledge gained and old knowledge discarded as obsolete. Therefore, in light of this fundamental transience, it seems quite apparent that the entire notion of "fact" based strictly on knowledge acquisition that is limited to our perceptual lens may appear naïve at best, if not egocentric. This is especially evident when we consider the vastness of universal spectrums, many of which are beyond the scope of our human understanding. Nonetheless both facts, as pieces of information that are presented as having objective reality as well as truth, have an equal opportunity of being attained.

Having this equal opportunity to be absorbed at any moment, facts, as well as truth, stand at the proverbial door and knock, waiting to be greeted and to become the governor of our thought processes. In essence, we are required to watch and be discerning, with the hope of being able to assess the difference. However, in so doing, please take special note of the word "presented" in the definition of "fact." Facts are aspects of reality that are simply *presented* in a way that makes them *appear* true. Lawyers, as well as countless others, often exploit this ability to frame the facts in a way that benefits their particular side of the argument. Truth, however, is backed by principles, enduring patterns, and a faithful consistency that transcends fact. Due to scientific reliance on empirical methods of research, however, as well as its emphasis on the quantitative rather than on the qualitative, we have come to consider the truth as tied to what can be seen with the eye or through an amplifying lens. We need proof of facts that may provide validity to the things that we will esteem as true.

How have we obtained these facts? According to many scientific studies, particularly those coming after the preliminary investigations that were conducted by Galileo, Newton, and Einstein, the evidence that something exists is most often preceded by noticing that something is missing. Based on an understanding of the laws of conservation, this was the way that the anti-particle of the electron—the positron—was first discovered as it accounted for the missing energy in the product of decay; as energy and mass needs to be conserved when work is done, any missing mass implies that it has simply been transferred to some other place by something.

In another way that focuses on something that is missing, many celestial bodies have been discovered by noticing their obvious effect on the motion of something else that exists in the visible realm. A vital point to remember is the *fact* that human limitations delineate the boundary of every experiment; all scientific discoveries remained unquestioned or in a conjectural state at one point in time. There was a time when they remained either in the realm of the unknown or in a lockbox phase, where due to an inability to break through a knowledge barrier wall, those discoveries remained, at best, in a potential state. During those phases, therefore, the elevated position of being labeled a fact couldn't be reached.

Paralleling the same reasoning track as the old tree in the forest allusion, was not all of this evidence always available? That scientists had not yet been capable of discerning these mass's existence did not in fact invalidate their existence. The point is that we cannot utilize the concept of something being a fact as a reason not to believe; facts are transient and facts often distort the pathway to the truth, the gaining of which does have benefit as John 20:29 contends: "*Thomas, because you have seen Me, you have believed. Blessed are those who have not seen and yet have believed.*" Blessing belongs to those who need not rely on their eyes to validate a truth.

A particularly interesting witness to the truth is demonstrated when people with a conscience are given a lie detector test; the truth is made evident because the sympathetic nervous system (something that is invisible) is vitally aware of the individual's physical, psychological, and emotional experience. As a vital witness of these experiences, this system cannot lie against that knowledge which resides, like a testament to the truth, within the nerves. The act of lying, going against what the nerves know to be true, causes stressful reactions that are measurable, and it's the consequent vibrational movement within this system that provides the evidence of the truth. Facts, however, can be manipulated. I'm reminded of something I once heard. It was suggested that the word "FEAR" can be accepted as an acronym standing for "**F**alse **E**vidence **A**ppearing **R**eal." And while this is sometimes an accurate assessment of FEAR, a similar acronym may be applied to "FACT"—"**F**alse **A**uthority **C**ountering **T**ruth."

Most of us live our lives succumbing to facts that are presented to us as real; understanding them as real, we stand under them and examine them from beneath. But the children of God have available to them the blessing of God that says among other things that "*the Lord shall make thee the head, and not the tail; and thou shalt be above only, and thou shalt not be beneath; if that thou hearken unto the commandments of the Lord thy God, which I command thee this day, to observe and to do them*" (Deuteronomy 28:13 KJV). Facts, therefore, are tied inextricably to a position in which people are in bondage to one thing or another that keeps them down, not allowing them to rise to a higher level.

While all facts cannot definitively be said to counter truth, those facts which align with physical evidence that we have come to interpret as unmovable are certainly suspect. Rightly considered to be lies, these *facts* place reasonable limits on what God is able to do. In his own reasonability, man projects his own ability onto God's, and if he feels incapable of affecting a particular end, the reasonable conclusion is that it is impossible. But Luke 18:27 declares a truth: "*The things which are impossible with men are possible with God,*" and Mark 9:23 encourages, "*If you can believe, all things are possible to him who believes.*" One's own belief in God's expansive capabilities, despite the facts at hand, is truly the instrument that tears down the walls of limitation; likewise it is one's own unequivocal trust in facts, as providing a map of reality, that constructs the defensive wall against hope and God's unlimited potential.

Truth, however, is like the gold in the Federal Reserve Bank; it provides solid security because it is not subject to decay. And as the gold reserves proclaim the actual wealth of a nation, so too does the level of truth upon which a people or nation is built. That which has no substantial reserve, which is backed by the standard of gold or some other precious substance, is therefore equivalent to an impoverished nation, or one whose foundation is built on lies. But what are lies? Lies are facts that have been presented in a twisted way, with the purpose of deceiving the hearer. First John clarifies where the root force of lies may be found: "*Who is a liar but he who denies that Jesus is the Christ? He is antichrist who denies the Father and the Son*" (1 John 2:22). Whenever Jesus Christ is denied, therefore, we know that that platform has been built on a foundation of lies.

Equivalent to both the Holy Spirit as well as the Word of God, as proclaimed in 1 John 5:6 and John 17:17 respectively, truth is equivalent to a treasure for which a man will sell all to obtain; it is as a "*treasure in the heavens that does not fail, where no thief approaches nor moth destroys*" (Luke 12:33). Accordingly, if the Holy Spirit and the Word of God have been proclaimed as equivalent to truth, then these two aspects of God, which have been provided to those who believe, provide not only a substantial and valuable treasure, but also a sure foundation.

First Corinthians speaks about this foundation and the role that it serves: *"According to the grace of God which was given to me, as a wise master builder I have laid the foundation, and another builds on it. But let each one take heed how he builds on it. For no other foundation can anyone lay than that which is laid, which is Jesus Christ. Now if anyone builds on this foundation with gold, silver, precious stones, wood, hay, straw, each one's work will become clear; for the Day will declare it, because it will be revealed by fire"* (1 Corinthians 3:10-13). As Jesus's alternate identity is the Word of God, we understand this foundation to be the Word as well as the gospel of Christ that is revealed therein.

As the writer of Corinthians, Paul alludes to himself as a master builder, but he isn't alone in building; all followers of the Word are likewise involved with their own construction, ostensibly upon the same foundation. What is implicit in this Scripture is a warning: While everyone builds, the product of one's building project is contingent upon the materials used. And in light of the fact that silver, gold, and precious stones are mentioned, we may rightly interpret these materials as relating to those things that the heart of man values, "*Lay not up for yourselves treasures upon earth, where moth and rust doth corrupt, and where thieves break through and steal: but lay up for yourselves treasures in heaven, where neither moth nor rust doth corrupt, and where thieves do not break through nor steal: for where your treasure is, there will your heart be also"* (Matthew 6:19-21 KJV).

As God esteems a man according to what his heart values, as 1 Samuel 16:7 alludes, "*For the Lord does not see as man sees; for man looks at the outward appearance, but the Lord looks at the heart,*" we may understand these materials as representing heart qualities gained through one's walk with the Lord. Whatever we esteem as precious is thus accounted as our treasure.

While every man's treasure will differ, a heart that is not deeply rooted in the principles of truth, such as godly wisdom and love, will more readily burn up when tested and tried. Those things that simply reside on the surface, having no root, when tested by fire will not withstand the flames; these things are likened to wood, hay, and straw. Those qualities that have taken root, however, and are firmly planted in the heart, may be likened to gold, silver, and precious stones, which when tried by fire will either be purified, or simply absorb the heat and remain intact. First Corinthians expresses what occurs after the foundation of the gospel of Christ has been built upon: "*If any man's work abide which he hath built thereupon, he shall receive a reward. If any man's work shall be burned, he shall suffer loss: but he himself shall be saved; yet so as by fire. Know ye not that ye are the temple of God, and that the Spirit of God dwelleth in you?"* (1

Corinthians 3:14-16 KJV). When people's ways emulate the ways of God in all sincerity and truth, it is as if gold, silver, and precious stones, which will withstand the flames of fire, have been carefully laid upon the foundation of the gospel of Jesus Christ.

This is well-pleasing to the Lord, and will ultimately glorify God, as it is His will that His children mature, grow, and function according to His plans and purposes. Utilizing each person's individual gifts, talents, and anointing, He will cause them to grow into the complete image of the a mature body of Christ with Jesus, as each member's governor, the head of the body: "*From whom the whole body, joined and knit together by what every joint supplies, according to the effective working by which every part does its share, causes growth of the body for the edifying of itself in love*" (Ephesians 4:16). God is glorified when His children attain their individually preordained potential as members of the body of Christ, which will enable every member to function under, and according to, the mind and will of Jesus Christ.

This spiritual building that involves each member of the body of Christ coming together and fulfilling the functionality of the office of Christ, after all, is the pattern of the building of Solomon's temple. While the physically constructed temple was made with hands, decorated with silver, gold, and precious stones, the body of Christ is a temple made without hands as it says in Acts 7:48-49: "*Howbeit the most High dwelleth not in temples made with hands; as saith the prophet, Heaven is my throne, and earth is my footstool: what house will ye build me? saith the Lord: or what is the place of my rest?*" (KJV). The true building of the temple of God, therefore, is not constructed out of visible, tangible building materials; it is rather a heavenly pattern where believers join together to construct the body of Christ, with the "*mind of Christ*" (1 Corinthians 2:16) as their head, governing each member according to their individual office. The effect of this oneness? Peace; and this peace enables God's rest.

Following the heavenly pattern of God resting on the seventh day according to Genesis 2:1-2 when He had finished His work, Jesus also finished His work when He was crucified on the cross. Knowing that He would rise the third day and complete the work that He had been predestined to complete, after receiving vinegar that had filled a sponge and been put upon a hyssop, while hanging on the cross, Jesus said likewise, "*It is finished*" (John 19:30). Both God's work as well as Jesus' at these moments marked completion of the laying of the foundation that would enable rest to follow, allowing those who believe to likewise enter into God's rest: "*There remains therefore a rest for the people of God. For he who has entered His rest has himself also ceased from his works as God did from His. Let us therefore be diligent to enter that rest, lest anyone fall according to the same example of disobedience. For the word of God is living and powerful, and sharper than any two-edged sword, piercing even to the division of soul and spirit, and of joints and marrow, and is a discerner of the thoughts and intents of the heart"* (Hebrews 4:9-12).

Things that are true, therefore, will adhere to heavenly patterns that are eternal because they exist in an unbound form, like vapor, and establish the foundational patterns of all that can and will be made on earth; in this regard, they would be considered a "first" creation. Things that are considered to be facts, however, align only with the substance of the second creation, or that which is made by hands, which, taking on a more solid form, proves to be more subject to decay—except, that is, if the solid is silver or gold. Accordingly, *"Thy kingdom come, Thy will be done in earth, as it is in heaven"* (Matthew 6:10 KJV) issues forth God's will that, according to the completed pattern in heavenly places, so shall it be on earth.

Chapter 6
Resonant Structures

As Moses was warned by God before he built the tabernacle, "*See that you make all things according to the pattern shown you on the mountain*" (Hebrews 8:5), *all* visions received by men reflect patterns that exist on an ethereal plane that is apprehended by the mind. While Plato was the first to stumble upon this understanding with his theory of ideal forms, this notion, far from glorifying man's intellectual abilities, highlights instead a realm that exists beyond man that enables visions and patterns to be received by him. Accepting such an understanding, however, requires one thing—humility. This notion implies unequivocally that man is not the originator of knowledge, but is as a steward into whose hands knowledge has been placed.

Acceptance of this idea, therefore, underpins man's ultimate reliance on a vision giver. And in light of Proverbs 29:18, which says, "*Where there is no vision, the people perish*" (KJV), man's very survival is on the line. If he is not provided with a vision of where he should go and what he should do, he remains stagnant instead. This vision then behaves as a principle of truth that if followed will provide men with sustenance. The sustenance, in turn, may be considered as an element of *fact*, which, while substantial, is nonetheless dependent upon the vision. Such is everything in heaven and earth. Heaven is the realm wherein truths abide and provide the enduring patterns seen and experienced in earth, and earth is the realm wherein facts that illustrate the eternal truths are verified. The visions that become guideposts for man, therefore, either glorify God or glorify man, whereby a spirit of pride, which wars against God, takes hold. These guideposts, in turn, can be obeyed or disobeyed, but may be understood as coming from outside of oneself in the form of frequencies that travel in the air.

As a radio that is set on a particular channel receives the information bound to its signals, so too are we. What we have in our hearts in abundance determines the channel that we are set on. When we are prepared to listen, therefore, the frequencies that travel on our set channel will be received as well as heard; they will cause a harmonious vibration within us, because of the agreement. When we have the Word of God in abundance in our hearts, we will resonate with all that is in agreement with it. And when there is no agreement, the frequency is not received, but resisted instead. First John 5:7 speaks about the Father, the Word, and the Holy Ghost as being one and being the record bearers in heaven.

Separate and distinct, albeit one—the Father, the Son/Word, and the Holy Spirit resonate with the record born by the other two; for instance when the Holy Spirit, who resides in believers, hears the Word, there is a similar vibration within the body to that which emanates from the Word of God; the Spirit resonates with the Word of God.

In addition, as all entities that bear records for a government function to see that the truth is upheld, so too does the Father, the Son/Word, and the Holy Spirit. These record bearers confirm that the Word is true, as the truth resides as a pattern in the heavenly realm. And as witnesses also testify in a court of law and provide substance to any claim, so too do the witnesses: "*And there are three that bear witness in earth, the Spirit, and the water, and the blood: and these three agree in one*" (1 John 5:8 KJV). As with frequencies that travel in the air and are enabled to be received by any and all who resonate with those frequencies, so too is it with those that bear witness to the truth in the earth. As witnesses, the Spirit, the water, and the blood not only resonate with the truth, but are also electrically conductive fluids that flow through the substances of the earth, by which the Spirit Himself might gain access. For one, blood is a conductor of electronic signals that are stored in low-frequency modes of vibration in the DNA, and for another, as discussed, electrolytes that conduct electric signals throughout a body flow in salt water. These physical substances and their properties thus represent the material facts that emulate a heavenly pattern, which is the conductivity of the Holy Spirit throughout the whole earth.

While water's property of electrical conductivity has been widely understood for decades, the conductive property of the DNA double helix had not been thoroughly investigated until Barton conducted experiments between 1993 and 1997. Typically involving both an electron donator and an electron receptor, which were separated along some distance on a strand of DNA, it was found that when light waves radiated an acceptor, a vibrating response or excitation would take place with the consequence that an electron current flowed toward the receptor from the donator.[4] Interestingly enough, this appears to mirror the heavenly pattern of the relationship between God and man. As a Donator, God's response toward us is contingent on the activity of our own minds and hearts, which evidence an excitation in response to His presence. As acceptors, that response within us incites an electric flow from God. Basically, the relationship is up to us, as James 4:8 proclaims: *"Draw near to God and He will draw near to you."* This is exactly what occurs in God's domain; when we think on His name, Jesus, or when we become excited by His Word, which is referred to as "*the light of men*" in John 1:4, that excitement then elicits a response from God, whose love is tangible.

4. Kuo-Chen Chou, Low Frequency-Resonance and Cooperativity of Hemoglobin, *Trends in Biochemical Science*, June 1989, http://www.sciencedirect.com/science/article/pii/0968000489900261.

The true significance of these experimental findings, therefore, is that the connection does not appear to be the Donator's (God's) responsibility, but rather the connection is contingent upon the receptor's ability to resonate with the things pertaining to God, which implies that a receptor already exists within those who receive. Several things of interest were found with these experiments, one of them being that according to reasonable estimations, the conductivity should have diminished with distance. However, according to experimental findings, this was not the case; distance did not appear to be an issue, either in terms of the conductivity of electrons between these two ports or in limiting the DNA double helix involvement in chemical reactions at a distance.

It would appear then that this piece of evidence actually validates the power of prayer, where the distance between acceptor and Donator is of no real consequence. In addition, when a chemical group was artificially bound to one end of the DNA molecule, the DNA was found to actually mediate a chemical change that occurred far down the helix, which was apparently responsible for mending a patch of damaged DNA. Herein lies a scientifically verified means by which "miraculous" healings actually may occur on a scientific level.

These findings support scriptural descriptions of Jesus speaking the Word as a Donator of power, and having it received by receptors who underwent physical transformations at the DNA level, even at a distance. Matthew 8:7-8 provides an example. After Jesus had told a centurion who had come to Him for help with his paralyzed servant, *"'I will come and heal him.' The centurion answered and said, 'Lord, I am not worthy that You should come under my roof. But only speak a word, and my servant will be healed.'"* The distance between Jesus and the man's servant didn't make any difference at all. Verse 13, however, makes a proclamation that stands as the prime determinant of the ability for time and space to collapse in this context: *"Then Jesus said to the centurion, 'Go your way; and as you have believed, so let it be done to you.' And his servant was healed that same hour"* (Matthew 8:13). The manifestation of God's power is thus contingent upon the receptivity of the receivers of the Word and power, with faith behaving as the bellows that provide the wind that initiates movement.

Add this to another finding, which was that as a DNA receptor instigates a response from the donator after it has been radiated, it becomes apparent that when white light (the Word) is actually received, something similar to photographic exposure occurs, where what is inside the heart (or blood) is revealed. This process appears to have been cited in Genesis 4:9-10 when Abel's blood cried out to God from the ground after he had been slain by his brother Cain: *"Then the Lord said to Cain, 'Where is Abel your brother?' He said, 'I do not know. Am I my brother's keeper?' And He said, 'What have you done? The voice of your brother's blood cries out to Me from the ground.'"* The DNA within Abel's blood not only pulsated with signals, but also flowed without resistance within the conductive fluid. Travelling as an electrical signal, the frequency within Abel communicated with God and transmitted a full account of what had occurred with Cain, which simply verifies Hebrews 4:13, which says, "*All things are naked and opened unto the eyes of him with whom we have to do*" (KJV). As with salt water, the Spirit of God flows through the blood and uncovers all secrets.

The idea that blood might "cry out" may seem not to fit with this due to the unfamiliar personification. However, if we consider the occurrence of sensory couplings such as what occurs with synesthesia, where the stimulation of one sense, such as sight, elicits an automatic response from another sense, such as sound, it may be more readily understood. While considered a neurological disorder, synesthesia nonetheless describes the potential for these phenomena to occur or rather to be perceived simultaneously. While sound and visible light register as different wavelengths on the electromagnetic continuum, it would appear that if lightning's effect on the air is said to result in a thunderous sound, then sound and light are conceivably coupled, albeit travelling at different speeds. Reasonably then, we may understand the DNA within the electrolytic fluid of blood in a similar way. When the receptor in a strand of DNA is stimulated by light, the light causes a similar vibrational response and registers as sound, perhaps not to us, but at least to God.

With regards to Cain and Abel's situation, Abel had been a shepherd and had pleased God with an offering of the firstborn of his flock; Cain, on the other hand, was a farmer and brought some of his fruit for an offering, but it apparently was not his best, for the Lord was not pleased. Cain became angry and resentful and killed Abel in response to his own feelings of apparent shame and inferiority. Here is an example of how bitterness and corresponding thoughts plant seeds in our hearts that can receive frequencies that guide us in the direction of our imaginations, where our thoughts don't align with God's and where our mouths run rampant. Such was the case for Cain. Yes, God had told him that he was cursed from the earth, that the ground would no longer yield its strength for him, and that he would be as a fugitive and vagabond, but the imaginations of Cain's heart planted a seed that received a frequency on which a worse report traveled; and Cain spoke a greater curse into being. He responded to God: "*Surely You have driven me out this day from the face of the ground; I shall be hidden from Your face; I shall be a fugitive and a vagabond on the earth, and it will happen that anyone who finds me will kill me*" (Genesis 4:14). He put a bounty on his own head.

The point is that through the initial avenue of thoughts, a connection was made with a spiritual frequency that promoted the dissemination of words that solidified the direction in which he was heading. Please understand that it is by habitually hearing and validating, as fact, the words and reports of those who invalidate the Word of God that seeds of unbelief are sown in our hearts. And as receivers, who ultimately respond to frequencies with which we resonate or agree, the implicit truth is that a similar frequency already resides within us. Therefore, what we hear and consequently believe as true is of paramount importance. The words that we allow to take root in our heart determine our reception of the frequencies that exist in the heavenly realm or, in other words, in the realm of the atmosphere.

Luke 8:18 warns, "*Therefore take heed how you hear. For whoever has, to him more will be given; and whoever does not have, even what he seems to have will be taken from him.*" "How we hear" in this context means the level of trust placed in what is heard. Whatever message has taken root, determined to be true, will resonate only with those who are like-minded and will consequently tend to resist all contrary messages. Perhaps this is another reason why Paul's warning in Philippians 2:5 should be heeded, which says, *"Let this mind be in you which was also in Christ Jesus."* As thoughts govern the mind and words make up thoughts, we can see that the level of knowledge and understanding is directly connected to the frequencies in the air to which we respond. The warning in Luke 8:18 exposes a truth: He who has the Word of God will connect with frequencies that will impart more godly knowledge; however, he who doesn't have the Word rooted in his heart will have any unsecured Word taken from him, because he will not be capable of connecting to the Word of God's frequency. Therefore, what is given is dependent upon what one already has.

Ephesians 6:12 alludes to all frequencies other than the one which connects with the Spirit of God and the Word of God. These are they with which we wrestle: "*The rulers of the darkness of this age, against spiritual hosts of wickedness in the heavenly places,*" and armed with this understanding, we may more fully realize our vulnerability. Needless to say, there are many opportunities to connect with frequencies of a less than desirable nature. All of these are potential donators that seek receivers whose hearts have been prepared to resonate with their "light." Matthew 6:22-23 expounds upon this particular light: *"The light of the body is the eye: if therefore thine eye be single, thy whole body shall be full of light. But if thine eye be evil, thy whole body shall be full of darkness. If therefore the light that is in thee be darkness, how great is that darkness!"* (KJV).

The light thus indicates the frequency or spectral wavelength that has been received by us. And as it is the electrons that flow at particular frequencies—the proposed thoughts that flow from either the Word of God, or from these rulers of darkness or spiritual wickedness—behave as seeds when they are received. When the seeds take root and the thoughts are consequently believed to reflect the truth, all incoming thoughts will subsequently cause resonance, resistance, or refusal of all subsequent incoming frequencies. Those that cause resonance will likely be further watered by words that correspond with that frequency; those that are resisted will likely be quenched, and those that are refused will likely be buried. Consequently, when we hear the Word of God and resonate with it, we are given an opportunity to further water the seed that was sown in our heart. We are also, however, given an opportunity to neglect it.

The Word of God, therefore, is like a seed (see Luke 8:11), which when planted in good soil near the sun transforms into the full vision of its DNA pattern. Just think about buying a package of seeds. When we purchase that package, we can see our garden transformed; and when we plant those seeds, we are fully persuaded that the seeds will yield the intended fruit, vegetables, herbs, or flowers. We are fully persuaded! Should the seeds not yield the intended crop, we would storm the retailer and demand a refund. We would demand restitution, because we would assume that we had been cheated and deceived by the package as well as the retailer. Why would we make this assumption? Because seeds don't deceive! They follow the pattern laid out in their DNA, and produce after their kind.

The problem as well as the solution centers on the seeds that are planted in our heart and consequently behave as receptors for donating frequencies. Whatever words we hear and believe to be true, therefore, will set our channel receiver to receive only that which flows thereon. All other frequencies will be resisted or will only produce static, or rather confusion. If we have the Word of God in our hearts, therefore, we have the receptor, which will resonate with any frequency that corresponds with the Spirit of God. But if we have only the seed of the words of the world's reports in our hearts, we will experience resistance to the Word of God instead. And as Matthew 12:34 says, "*Out of the abundance of the heart the mouth speaks.*" The words that come out of our mouth will correspond with the frequencies that have been received, and they, in turn, will produce after their own kind.

The system remains the same; it doesn't change. That's why Matthew 17:20 mentions that if you only have faith the size of a tiny mustard seed, you can tell a mountain to move and it will obey the command. The size of the seed doesn't matter; it's rather the seed's full persuasion that it can do what it was designed to do, which is to manifest its inner pattern. In other words, it knows what it is, and in light of its full persuasion that it is a mustard seed, no frequency that would try to persuade it to believe that it was an acorn or anything else would be received and believed. Herein lies the full weight of why it's so important to *take heed how you hear*. The frequencies to which we respond are the frequencies that determine the words that come forth out of our mouths. These, in turn, emit self-similar frequencies into the air, which when emitted, are once again prone to be received by the hearers, which causes a cycle to occur.

In response to this cycle, a great and terrible truth about the power of words becomes evident: Words have an undeniable ability to secure or destroy a person's sense of identity. Who a person believes himself to be is primarily determined by the words he speaks to himself, which often are in response to what he has heard others say or what he imagines they think about him. This is one reason why 2 Corinthians 10:5 provides us with the following action that relates to a weapon available to us: "*Casting down imaginations, and every high thing that exalteth itself against the knowledge of God, and bringing into captivity every thought to the obedience of Christ*" (KJV). The words that we speak to ourselves from the wells of our imagination come from our connection with all of the frequencies other than the one on which the Word of God travels. The Word of God, however, travels on a frequency that transcends the wells of our imagination; as truth, it travels on a frequency that connects us to the heavenly pattern, where our true identity is made known: "*And if you are Christ's, then are you Abraham's seed, and heirs according to the promise*" (Galatians 3:29).

Those who rely on the Word of God and the name of Jesus have been planted in the earth as seed whose purpose is to resonate with the Word of God and His Spirit, which will undoubtedly affect their thoughts. Thoughts then affect the words that are issued forth out of the mouth, and these words behave as seeds, which not only connect with corresponding frequencies, but also issue forth the characteristic spectral wavelengths into the atmosphere. A cyclical process thus takes root as those who have the seed of the Word of God will resonate with corresponding frequencies; resonation will affect the thoughts and words that are issued forth, and the words will be planted in any heart that has been so prepared.

Chapter 7
Appointments

As every seed has within itself the DNA blueprint of what it will be when it reaches maturity, we may refer to this final stage as the "appointed time." If we acknowledge all physical matter as bound states that have less potential than the vaporous states, many of which we do not see, then we may acknowledge the supremacy of the invisible things of this world. If everything first exists in the heavenly realm in an ideal form as Plato proposed, the initial state would be a conceptual state, or in other words, not apparent in the visible realm. This would thus constitute a "first creation" stage, wherein the process of any creative pursuit begins in the mind. There is therefore substance to the idea that creation begins with a thought, which, if considered along phase transitional lines, might be readily accepted as vaporous. Accordingly, we may understand our thoughts as having greater potential than ever before believed, which supports the idea that invisible things actually spawned creation.

"*For the invisible things [thoughts/conceptions] of him from the creation of the world are clearly seen [visible], being understood by the things that are made [physically manifested], even his eternal power and Godhead; so that they [unbelievers] are without excuse: because that, when they knew God, they glorified him not as God, neither were thankful; but became vain in their imaginations, and their foolish heart was darkened. Professing themselves to be wise, they became fools*" (Romans 1:20-22 KJV). Here, the "invisible things" speak expressly of the thoughts that behave as seeds and begin to germinate, and this constitutes the phase of the "first creation." However, this is just the beginning, because an undetermined period that requires faith and patience is required before the time of the second creation's fulfillment, just as there is when one is pregnant and must endure nine months of potential discomfort.

The differentiation between these two creations can best be understood as the difference between things that are *created* versus things that are *made*. Those things that are created are mentally conceived and become solid blueprints or patterns, whereas those things that are *made* adhere to the patterns provided in the first creation, but have substance, function, and form. Accordingly, we may assume that every second creation is preceded by a first, where thought precedes action. Acknowledgment of these two distinct processes, however, has not been clear. For instance, in the first chapter of Genesis, the making or creating of the heavens and the earth—and everything that exists in both spheres—appears to have been arbitrarily assigned. However, if *created* refers to a first mental creation, and *made* refers to a physical manifestation, this would imply that a period in which both faith and patience were put to work had to have been implemented between these two stages.

While such a waiting period does not appear to be specifically outlined, Genesis 2 begins, *"Thus the heavens and the earth were finished, and all the host of them. And on the seventh day God ended his work which he had made; and he rested on the seventh day from all his work which he had made. And God blessed the seventh day, and sanctified it: because that in it he had rested from all his work which God created and made"* (Genesis 2:1-3). While God's rest may have signified this period of waiting, this chapter speaks about things that were either created or made, so one can only assume that those things which were created had not yet manifested, while those things that were made had. Genesis 1:26, however, does present an exception: "*Let Us make man in Our image, according to Our likeness."* This signifies not an actual formation of man at this time, but rather a conceptual one. Clarification of the apparent separation between these two creations occurs in Genesis 2:4, as it begins to proclaim that this entire seven day period had not yet actually been manifested in the natural.

According to Genesis 2:4-5: *"This is the history of the heavens and the earth when they were created, in the day that the Lord God made the earth and the heavens, **before** any plant of the field was in the earth and **before** any herb of the field had grown. For the Lord God had not caused it to rain on the earth, and there was no man to till the ground."* Wait a minute! Genesis 1:27 had said, *"So God created man in His own image; in the image of God He created him; male and female he created them."* Man had already been created! Why did God say that there was no man to till the ground? Genesis 2:7 clarifies: Man had not yet become a *living being*: *"And the Lord God formed man of the dust of the ground, and breathed into his nostrils the breath of life; and man became a living being."*

There appears to have been some lapse of time between Genesis 1 and Genesis 2. God spoke the heavens and the earth and everything therein into existence in Genesis 1, but many of these things apparently remained in first creation seed form, hidden in the ground until the day that God caused the seeds to be watered, *"But a mist went up from the earth and watered the whole face of the ground"* (Genesis 2:6). This shift in perspective, allowing for an indeterminate period between when the words of faith were spoken in Genesis 1 and when the manifestation of those words bore fruit, alluded to in Genesis 2, in turn, would perhaps allow for the 13.7 billion years that research has estimated to be the age of the earth.

God's way of operation, the kingdom's way, therefore, is initially conceptual: "*Where there is no vision, the people perish*" (Proverbs 29:18 KJV). People's very survival depends upon what they can see with their hearts or with their eyes of faith. Habakkuk 2:2-3 proclaims: "*Write the vision, and make it plain upon tables, that he may run that readeth it. For the vision is yet for an appointed time, but at the end it shall speak, and not lie: though it tarry, wait for it; because it will surely come, it will not tarry*" (KJV). Conception, therefore, begins in the heart and in the mind, and is watered by words—written, read, and spoken. Following is a period where the conception takes root and is watered by words that encourage it to physically appear. This describes the first creation; it is simply a conception that may come into full bloom if one is fully persuaded that an appointed time exists in which it will appear.

The full persuasion that is alluded to is not unlike the response generated by a receptor in a strand of DNA when light excites it. As the receptor's response to the light in turn elicits a response from the donator, so too does a fully persuaded, faithful response elicit a response from God, as it says in Hebrews 11:6: *"But without faith it is impossible to please him: for he that cometh to God must believe that he is, and that he is a rewarder of them that diligently seek him"* (KJV). Therefore, God responds to those who believe that He is everything that He has said that He is, but to those who don't believe, there is consequently no response. As a receptor's response must travel on a frequency that resonates with a donator, it becomes apparent that angels may in fact signify these frequencies. According to Psalms 103:20, which says, *"Bless the Lord, ye his angels, that excel in strength, that do his commandments, hearkening unto the voice of his word"* (KJV), it is God's angels who are responsible for connecting to the spoken Word of God. A reasonable conclusion therefore is that as God's angels respond to the Word of God, Satan's angels respond to all words that may be perceived as anti-God.

Regarding God's angels however, Hebrews 1:14 declares them to be "*ministering spirits, sent forth to minister for them who shall be heirs of salvation*" (KJV), so in essence, these angels may be considered of equivalent value to spirits as well as to frequencies that connect to the seeds of the Word of God planted within the hearts of those who "*have ears to hear.*" And while Satan's angels, which joined him in his prideful fall, respond to all thoughts and words of a disobedient, anti-Christian nature, and water these seeds with more of the same, God's angels respond to and water all obedient seeds that correspond with the Word of God. In addition, as frequencies are considered spectral wavelengths, these angels undoubtedly have a color component; with consideration of Hebrews 1:7 which refers to the angels as a flame of fire, I am drawn to conclude that the white, scarlet, blue, and purple colors that appear in a flame are not a coincidence.

Nonetheless, the thoughts that correspond with obedience are presented in Philippians 4:8, which says, *"Finally, brethren, whatsoever things are true, whatsoever things are honest, whatsoever things are just, whatsoever things are pure, whatsoever things are lovely, whatsoever things are of good report; if there be any virtue, if there be any praise, think on these things"* (KJV). Therefore, when our minds meditate and linger on things that cause upset, seeds are planted in our hearts that enable connection with frequencies/spirits that engender strife and cause unfruitful words to be disseminated, which take root in a heart and produce after their kind at an appointed time.

Having the natural ability to generate pictures or visions, thoughts and words will invariably become the substance of either faith or its opposite, fear. While according to Hebrews 11:1, *"Faith is the substance of things hoped for, the evidence of things not seen,"* fear is evidently the substance of things *not* hoped for and is likewise the evidence of things *not* seen. Both exist in a conceptual realm, in the realm of the first creation, and therefore both are the determiners of the second creation's fulfillment. This then provides validation of Hebrew 11:1; faith *is* substance, for it provides the pattern to which the second creation must adhere.

What is abundantly evident, however, is that while a first creation must precede a second creation, there appears to be no cap on the question of "how long is too long to wait?" No matter what, it seems the fulfillment of a first creation requires faith and patience. "*For the vision is yet for an appointed time; but at the end it will speak, and it will not lie. Though it tarries, wait for it; because it will surely come, it will not tarry*" (Habakkuk 2:3). Everything in God's kingdom appears to belong to this system, where the first creation brings forth the second and the second appears at a particular time, which upon introspection, is usually deemed to have been "perfect."

Whether we understand it or not at the time, *"To everything there is a season, a time for every purpose under heaven: a time to be born, and a time to die; a time to plant, and a time to pluck what is planted"* (Ecclesiastes 3:1-2). Hebrews 9:27-28 contends, *"And as it is appointed for men to die once, but after this the judgment, so Christ was offered once to bear the sins of many. To those who eagerly wait for Him He will appear a second time, apart from sin, for salvation."*

Jesus' death and resurrection, therefore, was also a seed planted that has waited with faith and patience for the appointed time, when "*the sign of the Son of Man will appear in heaven, and then all the tribes of the earth will mourn, and they will see the Son of Man coming on the clouds of heaven with power and great glory. And He will send His angels with a great sound of a trumpet, and they will gather together His elect from the four winds, from one end of heaven to the other*" (Matthew 24:30-31).

As truth aligns with heavenly patterns and is simply mirrored by material facts, so too are the first and second creations. The first creation is planted firmly in the realm where vision prevails; the second creation, however, signifies the "appointed time" as Habakkuk 2:3 proclaimed: *"For the vision is yet for an appointed time; but at the end it will speak, and it will not lie. Though it tarries, wait for it; because it will surely come, it will not tarry."* The first and second creations, therefore, appear to refer to the rains spoken about in Jeremiah 5:24, Hosea 6:2-3, and Joel 2:23 respectively: *"Let us now fear the Lord our God, who gives rain, both the former and the latter, in its season. He reserves for us the appointed weeks of the harvest"* (Jeremiah 5:24). The former rain alludes to the gift of the first creation process, which mirrors God's way of operating, which He had imparted to man when he was first conceived or *created.*

This time period constitutes a *first day* that is implicitly conveyed in Hosea 6:2: "*After two days He will revive us; on the third day He will raise us up, that we may live in His sight."* These days, in turn, appear to reference the point when man was quickened by the breath of God and became a living being in Genesis 2:7. Man had already been created in Genesis 1:26, but was at this time *made,* and thereby fulfilled the first creation by bringing to pass the conception and articulation of a creature that was made after the image of God. Like God, he would be able to bring first creations to fruition.

Regarding the former and latter rain, Joel 2 references a point when the lapse between first and second creation will have seemed to contract:

> *Be glad then, you children of Zion, and rejoice in the Lord your God; for He has given you the former rain faithfully, and He will cause the rain to come down for you—the former rain, and the latter rain in the first month. The threshing floors shall be full of wheat, and the vats shall overflow with new wine and oil. So I will restore to you the years that the swarming locust has eaten, the crawling locust, the consuming locust, and the chewing locust, My great army which I sent among you. You shall eat in plenty and be satisfied" (Joel 2:23-26).*

In one month's time, both seed will be sown, or words spoken regarding some end, and the fruit, or product of one's words, will appear. Amos 9 also speaks of this appointed time:

> *"Behold, the days are coming," says the Lord, "When the plowman shall overtake the reaper, and the treader of grapes him who sows seed; the mountains shall drip with sweet wine, and all the hills shall flow with it. I will bring back the captives of My people Israel; they shall build the*

waste cities and inhabit them; they shall plant vineyards and drink wine from them; they shall also make gardens and eat fruit from them. I will plant them in their land, and no longer shall they be pulled up from the land I have given them," says the Lord your God (Amos 9:13-15).

This particular first creation appears to have been fulfilled in 1948, when Israel declared establishment of an Israeli state, and in 1952 when the Law of Return and Law of Citizenship were enacted that provided every Jew in the world with the right to settle in Israel.

Chapter 8
"Let..."

*Then God said, '**Let** the earth bring forth the living creature according to its kind: cattle and creeping thing and beast of the earth, each according to its kind'; and it was so" (Genesis 1:24).* As the chapter in which the initial seed in the form of thought and word was disseminated, Genesis 1 is the chapter in which we find the word *let* in abundance. From "*Let there be light*" (Genesis 1:3) to "*Let Us make man in Our image, according to Our likeness*" (Genesis 1:26), we see that the planting of these initial seeds, which corresponds with the former rain mentioned earlier, entailed God speaking forth the images of what He had already conceived in His mind.

The implication of this former rain is that the initial seed had already taken a conceptual or vapor-like shape and form, and only required being watered by words that established God's fully persuaded intention. In the same way that water becomes ice, or a vapor assumes a solid form through the process of losing heat energy, the watering of words that occurs through emitting sound waves that correspond with a spirit or frequency deposits them on the initial seed, as in the process of deposition. Recall that such was the process by which "*the Word became flesh and dwelt among us, and we beheld His glory, the glory as of the only begotten of the Father, full of grace and truth,*" mentioned in John 1:14.

Interestingly enough, we may attribute this loosing of heat energy to God's command, "Let."

> *For the mystery of iniquity doth already work: only he who now letteth will let, until he be taken out of the way. And then shall that Wicked be revealed, whom the Lord shall consume with the spirit of his mouth, and shall destroy with the brightness of his coming: even him, whose coming is after the working of Satan with all power and signs and lying wonders, and with all deceivableness of unrighteousness in them that perish; because they received not the love of the truth, that they might be saved (2 Thessalonians 2:7-10 KJV).*

> *For as in the days before the flood, they were eating and drinking, marrying and giving in marriage, until the day that Noah entered the ark, and did not know until the flood came and took them all away, so also will the coming of the Son of Man be. Then two men will be in the*

> *field: one will be taken and the other left. Two women will be grinding at the mill: one will be taken and the other left. Watch therefore, for you do not know what hour your Lord is coming (Matthew 24:38-42).*

When we compare God's command, *"let"* in Genesis 1, with this reference of "*he who now lets*," what becomes evident is that "*he who now lets*" is he who is obedient to God's commands, just as Noah was and was consequently taken away from the approaching devastation. *"He who now lets"* is also he who through faith and through the Holy Spirit "looses" or "lets" free first creations from the domain of the conceptual. This particular "loosing," of course, can only occur while those obedient to God's Word are still on earth. Accordingly, it would appear that there is a verifiable correspondence between this *letting* and the *loosing* of heat energy that is responsible for transforming seeds of conception into physical manifestations.

Not only does the transition between vapor and solid require a *loosing* of heat energy, but according to Matthew 16:19, this loosing is mentioned as a key to the kingdom of heaven: "*And I will give you the keys of the kingdom of heaven, and whatever you bind on earth will be bound in heaven, and whatever you loose on earth will be loosed in heaven.*" The processes of binding and loosing, therefore, are vital components of God's way of operating. Accordingly, we may consider the act of speaking words that "call into being" those preconceived seeds, like the latter rain, loosing them from their present phase state. This also supports God's intention of creating and making man in His own image, after His likeness, for God is one who *"calleth those things which be not as though they were*" (Romans 4:17 KJV).

Chapter 9
Keys

A key issue in the kingdom of heaven aligns with the processes that govern all that appears in the earth. From phase transitions and quantum physics to biological metabolism and nuclear reactors, evolutionary passages from one state into another appear to be involved. From the binding of matter into a solid and the binding of quarks into nucleons to the loosing, or dividing, processes of mitosis and meiosis and the loosing process of fission, all of these processes involve either binding matter together or loosing the bound entities from a bond.

These processes, therefore, being evident on every level, confirm why Jesus regarded these processes as "*keys of the kingdom of heaven*" (Matthew 16:19). Speaking to Peter, Jesus indicated the importance of these processes by proclaiming, "*And I will give you the keys of the kingdom of heaven, and whatever you bind on earth will be bound in heaven, and whatever you loose on earth will be loosed in heaven*" (Matthew 16:19). While these processes appear to relate to most every process involved in life, one thing appears significant in light of one particular statement Jesus made: "*If you abide in My word, you are My disciples indeed. And you shall know the truth, and the truth shall make you free*" (John 8:31-32).

The concept of freedom then becomes of paramount importance, in that it appears to be a most desirable state. With the consideration that according to phase transitional processes, a solid phase represents the bound phase, and a vapor phase represents the loosed or free phase, achieving this latter state becomes an unspoken goal. But how does one become free? As a solid requires absorption of heat to transition into either a liquid or a vapor, this loosing also requires heat absorption. This absorption, however, may be understood as signifying the absorption of the Word of God. And in this way, we may more fully understand that if Jesus' disciples continued in His Word, His Word would make them free.

We must understand, however, that in His proclamation about the keys of the kingdom of heaven, Jesus gave no definitive value to either binding or loosing. Therefore, it would appear that while the free vaporous phase appears supreme, there is verifiable worth to a bound phase or to the process of binding. For instance, the process that enabled Jesus, the Word, to became flesh and dwell among men in John 1:14, refers to an exothermic process, whereby heat energy is released as opposed to absorbed. This release of heat energy, in turn, signifies a process that, contrary to leading to freedom, leads instead to a bound form or perhaps "bondage." However, it is through this process that solid things appear to materialize as from thin air. Therefore, we can summarize the value of these phases in this way: We may become free in mind, body, soul, and spirit when we continue in the Word of God; this then would constitute an absorption of heat energy, which enables a transformative operation on the *inside* of us. When we consider the value of *releasing* heat energy though, we may see this in the context of releasing words of faith, whereby their release enables a transformation from a vaporous phase into a solid phase, mirroring the process whereby the *Word became flesh* in John 1:14.

Romans 4:17 speaks of this precise release: "*even God, who quickeneth the dead, and calleth those things which be not as though they were*" (KJV). God conceptualizes what He desires, whereby He plants an initial seed as a first creation. Then He waters that seed with His Word, which brings forth the full manifestation of the conceptualization. In this process, He calls forth things that, because they exist only in conceptual form, remain hidden to the senses. However, with faith as the hydraulic force that issues forth His Word, the seeds of faith are thereby watered and enabled to grow; and according to the second creation's appointed time, the seeds must obey and the vision must materialize.

Not only did Jesus give Peter and all who believe in His name these keys of the kingdom of heaven, but as man has been created in God' image, after His likeness, these processes govern the ways in which man is to take his rightful dominion over the earth. As God brought forth the "*substance of things hoped for*," and provided "*the evidence of things not seen*" (Hebrews 11:1), man, who was made after God's image and likeness, occupies the same office. Interestingly enough, while the true use of these keys was purposed to bring forth seeds of faith to fruition, the counterfeit use of these processes appears to be forceful dominion, whereby man binds all who are beneath him, relegating them to the shadows, and looses himself from responsibility. Aligning with hatred or fear, we know that these processes counter God's ways, because, "*God is love*" (1 John 4:8), and love "*is not puffed up; does not behave rudely, does not seek its own, is not provoked, thinks no evil*" (1 Corinthians 13:4-5).

While the keys of the kingdom of heaven that Jesus gave to Peter are based on the processes of binding and loosing—"*whatsoever thou shalt bind on earth shall be bound in heaven: and whatsoever thou shalt loose on earth shall be loosed in heaven*" (Matthew 16:19 KJV)—what does this truly mean? First, the question of why these keys were given to Peter in the first place must be addressed. The apparent gifting of these keys was in direct response to Simon Peter providing a correct answer to the question of who *he* thought Jesus was.

"You are the Christ, the Son of the living God," answered Peter (Matthew 16:16). In verses 17-18, Jesus answered Peter's accurate reply: *"Blessed are you, Simon Bar-Jonah, for flesh and blood has not revealed this to you, but My Father who is in heaven. And I also say to you that you are Peter, and on this rock I will build My church, and the gates of Hades shall not prevail against it"* (Matthew 16:17-18). Peter's correct response was exhibiting to Jesus that Peter himself had joined with the Father and because of this binding with the Father, he had loosed himself from the bonds with men, with whom the social pressure to belong overtakes many.

Through the frequency of the Holy Spirit of God, Peter received revelation knowledge that Jesus was the Son of the living God, and having *heard* Jesus' Word, his own heart had been trained to align with God's truth. When Jesus consequently acknowledged that this transition had occurred, He saw the fulfillment of what had been envisioned as the foundation of Jesus' body, His church—the infilling of the Holy Spirit, which would come after Jesus had risen from the dead and would enable direct communication with God the Father. This enablement thus constituted the true *Rock* upon which Jesus would build His church. One vital process of binding and loosing, therefore, involves the forming and disintegration of connections; while this describes myriad physical processes that make up our universe, in essence, it also describes the forming of one's knowledge base, which requires, among other things, the acceptance and resistance of particular connections.

While spoken in an implicit way, the keys of the kingdom of heaven are clear and they allude to the fundamentals involved in the strong force, the weak force, chemical bonds, nuclear fission and fusion, as well as wave dynamics, to name just a few. The specifics involved are varied, but can generally be conveyed as the joining together or tearing apart of matter, which has both a particle and wave nature, a property that constitutes the wave-particle duality of matter.

Functioning at the microscopic level, two key processes that may be said to epitomize the processes of binding and loosing are fusion and fission, respectively. While both processes involve the release of energy, nuclear fusion is the fusing of two or more smaller atoms into a larger one. Nuclear fission, however, is the splitting of an atom into two or more smaller ones. With nuclear fusion, the nucleotides (protons and neutrons) are bound together by a nuclear force characterized by its strong binding energy. Nuclear fission, however, entails breaking the bond between the nucleotides and happens to be the root force that produces nuclear explosions. Consequently, in order for nuclear fission to take place, a negative binding or loosing energy must surpass in strength the positive binding energy, and must infiltrate the bond at an atomic level. The strength of fission, therefore, exceeds fusion's positive binding energy's strength.

Matthew 12:29 comes to mind: *"How can one enter into a strong man's house and plunder his goods, unless he first binds the strong man? And then he will plunder his house."* In order to break through anything strongly secured, one needs to overcome by being armed with a stronger weapon. As a consequence of this negative binding energy being greater than the positive binding energy, the bond between the nucleons is loosed, and the atom splits into two new nuclei and releases excess stored energy, often in the form of gamma radiation. And here we have a nuclear example of what is often required to be an overcomer: We must surpass the strength of the "strong man" and understand the explosiveness involved in trying to set the captives, if not just ourselves, free.

While the process of fission can occur spontaneously, it is often accomplished with the help of free neutrons, as in the case of nuclear reactors. In the condition that there is a large enough atomic nucleus that can undergo fission, such as uranium-235 or plutonium-239, these nuclei may readily absorb a neutron that will instigate a process of fission. According to the maxim, "the sum is greater than the parts," when nuclei that are lighter than iron fuse together, energy is lost in the binding process, but the final nucleus is characterized by a greater binding energy. Accordingly, it might appear that the lost energy that occurs in the process of binding the nuclei together is actually stored in a potential form; for example, while it is undetected and appears to be missing, it has simply transitioned into an invisible force that will potentially reveal itself as an explosion of light should loosing occur. In this way, the process of binding appears to correspond to the initial seeding process, or to the *former rain,* and the process of loosing appears to correspond to the *latter rain* or manifesting stage. Light, therefore, may be understood as a manifestation of potential energy.

Chapter 10
Light's Duality

According to the understanding that Genesis 1 spoke of things that had not yet fully materialized, but had been primarily created or conceptualized as patterns of what would come to pass, light was first conceived before it was brought forth. It was therefore in this conceptual stage that we can truly understand light as a potential energy. However, when God said, "*Let there be light,*" and light *was*, light apparently manifested within a presumably measurable period. Nonetheless, like everything else that God created and made, before light was *let* loose, with explosive power as by a fission event, it had remained bound as in a seed.

While we may regard light as a manifested potential, or as a second creation, the property of wave-particle duality appears to extend our notions of universal dynamics, where the keys of binding and loosing may be understood in new ways, and where these dynamics may provide reflections of heavenly patterns. While a particle of light is referred to as a photon, an energy carrier (boson) of the electromagnetic force, a light wave consists of oscillations that transfer energy (for instance, potential energy) from one point to another. While a photon can excite individual electrons within atoms and affect their position, a light wave involves all of the atoms within the wave. If we consider that the Word/Jesus is referred to as the "light of men," we must examine this phenomenon as illustrative of at least one way in which God operates.

If Jesus is light, therefore, His modes of functioning would be both as a particle as well as a wave. As a particle, He would have the ability to be sent forth into the world with the potential of being absorbed by an electron within an atom and causing excitement. That excitement, in turn, would allow the electron to jump up another level, and then, needing to get back to where it started, it would release the excess energy as light. Well, interestingly enough, this epitomizes what Jesus did! He was sent into the world so that He might enter into people's hearts and cause them to become excited; in response, they would emit a corresponding light out into the world. This also aligns with the resonance that occurs when frequencies find other frequencies with which they agree.

We see this potential emitting of light in Matthew 5:16: "*Let your light so shine before men, that they may see your good works and glorify your Father in heaven."* And once again, we have an example of light as having existed in a conceptual stage before it was actually brought forth; but in this example, the conceptual light existed *within* man. We may therefore conclude that these thought-inspired frequencies/spirits/angels or thoughts register on the electromagnetic spectrum as light, and may therefore assume particle as well as wave properties.

As a wave, however, light cannot be seen as having the sole mission of finding a frequency with which it may resonate. Unlike a particle, a wave is already characterized by a definitive level of agreement; in other words, a wave is not a wave unless neighboring particles are following the same track. As light can behave according to both of these dynamics, however, we may use this analogy to extend our understanding of the Father, the Son—Jesus Christ and the Word of God—and the Holy Spirit.

Alluding to Jesus, John 1:1-4 spoke about the Word, saying, "*In the beginning was the Word, and the Word was with God, and the Word was God. He was in the beginning with God. All things were made through Him, and without Him nothing was made that was made. In Him was life, and the life was the light of men."* Accordingly, it would appear that Jesus, the Word, who was referred to as the "light of men," behaved specifically as would a photon. As energy carriers, photons carry discrete packets of light that can be absorbed by electrons in atoms if the electrons have a matching frequency. If absorbed, the electrons become excited, potentially jumping up a level, and in the process of settling down, they will release excess energy in the form of light. As electrons may be considered to be equal to thoughts, this makes sense. If the light that is embedded in the Word of God, Jesus, is absorbed by thought particles or waves that are in agreement, there will be a consequent excitement that is represented, not only by an expansion of thought, represented by jumping to another level, but also by a subsequent emitting of light.

Referred to as the Word of God, Jesus emits the light that is embedded in the Word. When we admit that same frequency of light into our own hearts, if there is resonance, our thoughts (electrons) will change. This change then represents a transformation that occurs in response to the renewing of the mind, which is spoken of in Romans 12:2: *"And do not be conformed to this world, but be transformed by the renewing of your mind, that you may prove what is that good and acceptable and perfect will of God."* And in Philippians 2:5, we read, "*Let this mind be in you which was also in Christ Jesus,"* which for one, stresses that we can allow the same thoughts that governed Jesus to govern us.

Illustrating the conception stage of a seed by the use of the word *let,* as we saw in Genesis 1, "*Let this mind be in you,"* appears to be stressing that an opportunity exists wherein a conception that remains in a bound state of dormant potential can be released and *made* free and living. An example of such a release is found in Genesis 2:7 when God breathed into Adam's nostrils the breath of life and he became a living soul. This breath of God as well as the light that is embedded in the Word of God would both appear to enliven that which draws in its power.

While Jesus alone may be regarded as the particle form of light that fundamentally seeks out electrons or thoughts with which He may resonate, the Father and Holy Spirit may more readily be conceived as waveforms of light. As such, all of the dynamics that apply to waves likewise may find applicability. When God said, "*Let there be light, and there was light*" (Genesis 1:3), those words went forth as waves—sound waves. And as sound waves, words require matter through which to travel, and should they interact with other sound waves, the property of interference will be experienced. As atoms can interact with other atoms through basic attraction or repulsion, waves interact with each other through constructive or destructive interference, and while interacting atoms can either remain pure or become mixed, so too can waves.

As the keys of the kingdom of heaven are binding and loosing, when two waves interact and become bound to one another or if they are bound and become loosed or freed, it may be understood as a binding together or a loosening of the wave's strength and direction. And as atoms have a particular identity when they are pure and another when they are mixed, if the strength and direction is definitive, so is the wave's "identity." As waves are oscillations with crests that reach into the air and troughs that fall beneath, the greater the reach (amplitude) or the fall, the more turbulent a wave may appear. In the event that waves constructively interfere with one another, the waves will be considered "in phase," meaning that their crests and their troughs line up when they interact, and the waves gain strength through a process where the waveforms bind together. In the event that waves destructively interfere, however, the waves will be considered "out of phase," with one wave's trough colliding with another wave's crest, and strength will be diminished (loosed).

Aligning with the keys of the kingdom of heaven, we see these interferences play out in human interactions, with energy either being gained (bound) or lost (loosed). As we have been created in the image of God, we have been made to emulate Him who *"calleth those things which be not as though they were"* (Romans 4:17 KJV), which strongly suggests that our words have power. And in actuality, our words function with the potential of positive binding energy (binding) as well as negative binding energy (loosing), and they can either constructively or destructively interfere with other spoken words, which are also emitted in conjunction with nonverbal frequencies with which the words may agree or not.

Note that at this point in space and time, there is no intrinsic value to either binding or loosing, or constructive or destructive interference; their value must instead be assessed contextually. For instance, with regards to human interactions, when we interact with someone who has a similar interpretation of reality as we do, perceiving things in the same fundamental way, our waves are "in phase," and there will be a greater likelihood of excitement to occur. The amplitude of the interfering waves can be seen as having grown during the interaction. When, however, we interact with someone with whom we have nothing in common, we can feel the tangible effect of destructive interference. If we happen to have more energy, our crest will collide with their trough, with the result that the amplitude of our crest will be brought low, which is experientially felt by a draining of our energy.

The true importance of interfering factors, however, rests on the understanding that whatever we join and invariably bind ourselves to will gain strength as in constructive interference, and whenever we are not resonating with a frequency, there is the potential for destructive interference. However, we have equal responsibility in determining the quality of the interaction, as our words can be sent forth to stir up a wave or to settle it down. We can send forth frequencies that correspond with strength, peace, joy, and love or weakness, unrest, sorrow, and hatred.

While both the keys of the kingdom, binding and loosing, can be understood as being applicable to many natural processes, the process of loosing may be regarded as having greater intrinsic value. It is, after all, the process that brings forth the full manifestation of a seed, as does giving birth. But, the process of binding is equally valuable, for this process is involved in shaping the initial seed of faith, which, in the form of its conception, is even smaller than a mustard seed; it abides in the realm of the invisible. But as Zechariah 4:10 reminds us: *"For who hath despised the day of small things? for they shall rejoice, and shall see the plummet in the hand of Zerubbabel with those seven; they are the eyes of the Lord, which run to and fro through the whole earth"* (KJV). A conception is small, but it is a beginning with a sure end, as Habakkuk 2:3 contends: *"For the vision is yet for an appointed time; but at the end it will speak, and it will not lie. Though it tarries, wait for it; because it will surely come, it will not tarry."*

Chapter 11
Seeds of Faith

As a seed of vision, the things that were not, which God called as if they were, were first without form and void, similar to what is spoken about the condition of earth in Genesis 1:2: "*The earth was without form, and void; and darkness was on the face of the deep*." While the earth here is said to be without form and void, the fact that it "*was*" is not without significance. It, therefore, existed, but it was apparently not available. It sojourned in a phase space that was dark and empty, and in this way, we may understand that the earth was bound by darkness and was therefore simply awaiting freedom and being loosed from its dark confinement. In essence then, we may consider this phase transition as an illustration of what occurs in the process of "overcoming."

While First John mentions overcoming in the context of he who overcomes, "*For whatever is born of God overcomes the world. And this is the victory that has overcome the world—our faith*" (1 John 5:4), overcoming is actually epitomized in the process of loosing. With consideration of nuclear power, the process of fission or loosing results in explosive power. And in the example of giving birth, again fission occurs with explosive power. When a seed bursts forth and the delicate sprouts break through floorboards or concrete, we understand the power of loosing. Loosing is the process that enables a *breakthrough to* occur. Experientially, we may validate this by acknowledging the mentally bound state of stagnation that usually exists prior to a breakthrough moment, when suddenly one sees what they could not see before.

If we were to imagine a seed as a tightly bound entity whose sole purpose was to one day be loosed so that it could fulfill its potential, then we might perceive a key operative of the kingdom of heaven in its true light. Nonetheless, while positive binding energy is the glue of all matter, negative binding energy *overcomes* this glue. So when Jesus gave Peter the keys of the kingdom, He was communicating that by providing His Word to His disciples, they were fully equipped to loosen the bonds that were keeping the multitudes bound. At the same time, the Word would create new things, new evidences that would appear after the loosing or splitting/dividing of the initial "heavy" nucleus (or heart). And it is with consideration of this particular process wherein fission's importance becomes truly applicable.

There appears to be a connection between the Word of God and the negative binding energy or loosing involved in fission: Both overcome the bound. Considering the keys of binding and loosing that were given to Peter, as well as the scriptural references to how the words that we speak can both bless and curse, "*But no man can tame the tongue. It is an unruly evil, full of deadly poison. With it we bless our God and Father, and with it we curse men, who have been made in the similitude of God. Out of the same mouth proceed blessing and cursing*" (James 3:7-10), it would seem fitting that words would be stored in a neutral vessel.

Interestingly enough, neutrons happen to be neutral vessels, with no definitively positive or negative charge, with which it may attract or repel other particles. More importantly, however, is that neutrons also have been touted as being the key to nuclear power. Nuclear power, in turn, may be understood as the energy that is released or loosed from a bound state. But something must break through the bound state in order to initiate the breaking apart of the bound state. This is where neutrons come in. Having no charge, they do not readily experience resistance, and can therefore be absorbed relatively easily if they are travelling at a slow enough speed. When one is absorbed from a heavy enough nucleus such as uranium, this absorption initiates a process of loosing or dividing the nucleus into smaller nuclei. And when the loosing or splitting occurs, more free neutrons are expelled as well as photons in the form of gamma radiation. The free neutrons that are emitted can then be absorbed by other nuclei, and a chain reaction ensues.

This is significant, because this process of loosing or fission is patterned after the image of a breakthrough. Similarly requiring something that will loosen from a bound, stagnant state, such as darkness, a breakthrough thus requires something that is patterned after a free neutron. Not bound within a nucleus, a free neutron travels at speeds that can readily cause other nuclei to heat up and break apart. The broken-up hot material then initiates either ionizing or non-ionizing radiation, such as radio waves, light, and heat that either behave as conveyers of information or as electron liberators (liberators of thought). Taking into consideration the ionizing component, we become aware that radiation, like most things, must have its value assessed contextually. A free neutron, therefore, provides key information into what may cause a breakthrough to occur.

Proclaimed to be a key to nuclear power, it does seem curious to me that the neutron has been considered a fermion rather than a boson. While I wish not to question the spin-statistics theorem that relates the spin of a particle to the particle statistics it obeys, I might suggest that neutrons do not operate strictly as fermions. While the Bose-Einstein statistics wisely pose no limitations on the quantum states of particles (bosons or force carriers) to which the statistics apply, the limitations posed by the Pauli exclusion principle—which fermions, which are associated with matter, are said to obey—appear to have been short-sighted.

The Pauli Exclusion principle appears suspect, considering the recent measurement of both a particle's position as well as its momentum, thereby invalidating Heisenberg's uncertainty principle, which had stated the impossibility of this simultaneous measurement. Denying the simultaneous occupation of quantum states, it too may be doomed to failure. In light of both principles' reliance on perceptual measurements, it seems likely that any principle based on a presumably limited capacity for measurement may not endure for very long. That human beings may not have the capacity at any one time to measure a quantity neither makes it a fact nor a falsehood. If Heisenberg's principle was shown to be inaccurate, other scientific principles upon which assumptions are based may be wisely reopened for further investigation. Interestingly enough, one of the greatest scientists of all times, Albert Einstein, responded to Heisenberg's proposal of the uncertainty principle, saying, "God does not play dice."

That the neutron may inhabit a mediating position between an energy-carrying boson and a matter-forming fermion appears not to be out of the realm of possibility. Despite the composite nature of neutrons, being made up of quarks, as well as their intimate involvement in the production of matter, if a neutron could be considered a "word-carrier," it would actually satisfy both the Pauli exclusion principle as well as the Bose-Einstein statistics. As a mediating particle somewhere between a fermion and a boson, just as Jesus is a mediator between man and God, this identity would certainly validate the neutron's proposed identity as the key to nuclear power.

Considering the neutron as a fermion-boson—wherein it was actually capable of transitioning from one state into the other—would not only be an explanation of its neutrality, but would provide an answer for its ability to initiate a nuclear reaction. In alluding to the process of a chain reaction, James speaks about words and fire:

> *And the tongue is a fire, a world of iniquity. The tongue is so set among our members that it defiles the whole body, and sets on fire the course of nature; and it is set on fire by hell. For every kind of beast and bird, of reptile and creature of the sea, is tamed and has been tamed by mankind. But no man can tame the tongue. It is an unruly evil, full of deadly poison. With it we bless our God and Father, and with it we curse men, who have been made in the similitude of God. Out of the same mouth proceed blessing and cursing (James 3:6-10).*

This Scripture illuminates the nature as well as the power of words. As they are expressed and propagate through matter as sound waves, the interference factors are relatively obvious, but that they initiate and maintain chain reactions is a frightful but accurate assessment. Experientially, we all can validate this truth, as we can encourage ourselves with words and we can discourage ourselves with our own self-talk; other people can make our day with pleasant words or they can bring us down; it's palpable. We know it, we feel it, and we live it. Understanding it as equal in magnitude to nuclear power is just one step further into recognizing its daunting power.

While all words do have power, the Word of God is the only provider of substance that, like light, continues to create. Hebrews 11:1 illuminates this truth: "*Now faith is the substance of things hoped for, the evidence of things not seen.*" Therefore it is hope that enables the creation of an initial seed of a vision, which when watered with the Word of God secures its appointment; it has no option other than to materialize due to Romans 10:17, which accurately claims that "*faith comes by hearing, and hearing by the Word of God.*" So hearing, believing, and ultimately, speaking the Word of God initiates a chain reaction, just like what occurs when uranium absorbs a free neutron in a nuclear reactor. A neutron (the proposed word carrier) is absorbed and causes the uranium to undergo nuclear fission, splitting into lighter nuclei, and in the process, releases more free neutrons as well as high intensity electromagnetic waves.

Due to the fundamental process of absorption, which implies a full digestion, absorbing words involves, above all, *hearing* or rather listening with an intention of understanding what is being said. When words are absorbed according to this understanding, the simple fact that the words are understood implies that a receptor seed has already been planted. Having already been planted and presumably taken root, this receptor seed behaves like an electron in an atom that resides in a particular energy shell/level. In order for this electron to become stimulated and moved out of its stagnation, an incoming photon with a similar energy must strike it. Without an electron with a similar energy as the photon, however, the photon would have no place to rest, and it would pass on by and consider that atom transparent. Therefore, whatever words have been *heard* and accepted as true take root and develop as seeds that will then behave as receptors for words or frequencies with which they resonate.

In light of this analogy, we may readily consider John the Baptist's message of repentance as an initial seed, for he preached in the Judean wilderness, saying, *'Repent, for the kingdom of heaven is at hand!' For this is he who was spoken of by the prophet Isaiah, saying: 'The voice of one crying in the wilderness: "Prepare the way of the Lord; make His paths straight"''"* (Matthew 3:1-3). His water baptism and message of repentance was, therefore, given as a preparatory measure. People's subsequent obedience, interestingly enough, absolutely did plant a seed in their hearts, which then enabled them to receive Jesus' words as well as His Spirit when that became available.

In essence, then, we may understand the full implication of the words that we accept into our heart and believe as true. They become seed that will affect all subsequent absorption or resistance. "*He who has ears to hear*," therefore, speaks about those who have, intentionally or unintentionally, had the ground of their hearts prepared with a receptor seed. When the Word of God or His Spirit knocks, the receptor seed will resonate, as did John the Baptist inside of Elizabeth's womb when Mary came to visit her after she herself had conceived: "*And it happened, when Elizabeth heard the greeting of Mary, that the babe leaped in her womb; and Elizabeth was filled with the Holy Spirit. Then she spoke out with a loud voice and said, 'Blessed are you among women, and blessed is the fruit of your womb!'*" (Luke 1:41-42). He who *does not have ears to hear,* however, is he who has no receptor seed for the Word of God within his heart, and therefore cannot resonate with it. If the world's words and reports have been absorbed, the Word of God is likely to be resisted and the heart of whoever heard will resonate with any frequency or spirit that resonates with those words instead.

Chapter 12
Faith's Potential

God's way of operating in the world was, is, and will always be through faith, as He *"calleth those things which be not as though they were"* (Romans 4:17 KJV). He calls into existence the things that only exist in the invisible realm, in the realm of conceptual pattern. "*Now faith is the substance of things hoped for, the evidence of things not seen. For by it the elders obtained a good report.* [The good report already existed as a conceptual pattern in the invisible realm.] *Through faith we understand that the worlds were framed by the word of God, so that things which are seen were not made of things which do appear"* (Hebrews 11:1-3 KJV).

The seeds of faith spoken about in the last chapter specifically refer to this account of faith that both enables us to understand and is the provider of substance to those things which have been conceived by the mind. Regarding the subject of understanding, 1 Corinthians 2:14 states, "*But the natural man does not receive the things of the Spirit of God, for they are foolishness to him; nor can he know them, because they are spiritually discerned.*" Therefore, in order to recognize and mentally comprehend the things of God, one must have within them a discerner; one must have a receptor seed for the Word of God, and this receptor seed may be understood as the actual substance of things hoped for, or rather, faith.

That the operations or laws of God as seen in earth must correspond with those patterned in heaven appears to be the primary structural foundation of everything, as is faith. This is difficult to perceive due primarily to the human self-imposed limitation of reliance on sense knowledge. Sense knowledge, of course, relies primarily on sight and sound, which, when we consider the entire electromagnetic spectrum, constitutes a very narrow band of frequencies. Nonetheless, the heavenly realm consists of conceptual patterns that exist as seeds of faith that have the potential to be watered with faithful words. If consistently watered and not abandoned, there is an appointed time of fruition. But it all begins as an invisible seed, a conceptual seed that aligns with Colossians 1:15-16.

Conveying information about Jesus as the Word of God, the Scripture says: *"He is the image of the invisible God, the firstborn over all creation. For by Him all things were created that are in heaven and that are on earth, visible and invisible, whether thrones or dominions or principalities or powers. All things were created through Him and for Him"* (Colossians 1:15-16). As Jesus is the Word of God and is the image of the invisible God, we may conclude that the Word of God is invisible and travels on frequencies, as does sound and visible light, that communicate to God and His angels.

The earth's potential, therefore, can only be seen through a spiritual lens, or through the eyes of faith. In this respect, we may understand the earth's potential in heaven as existing in a vaporous phase. Accordingly, in order to fulfill a vapor's *potential* of transitioning into either a liquid or solid phase of matter that, according to gravitational laws, might abide on earth, a vapor would have to give up heat energy. As every earthly process emulates a conceptual preexisting pattern, which also parallels a heavenly pattern, an equivalent spiritual process must likewise apply.

If "*the Word became flesh"* in John 1:14, which alluded to the birth of Jesus, the Word was at that point a vapor and was deposited on a preordained egg in Mary's womb. Assuming that the Word was vaporous then, the process of deposition having occurred would suggest that the Word lost heat energy, and in order for that to occur, the Word would need to have been spoken. Alluding to and confirming this process of deposition as coming forth from the seemingly invisible, Isaiah 55:10-11 declares: *"For as the rain comes down, and the snow from heaven, and do not return there, but water the earth, and make it bring forth and bud, that it may give seed to the sower and bread to the eater, so shall My word be that goes forth from My mouth; it shall not return to Me void, but it shall accomplish what I please, and it shall prosper in the thing for which I sent it."* The Word of God, like rain and snow, was deposited upon a preexistent substrate of matter and brought forth His/its own intentions, which correspond with matter's DNA pattern.

This Scripture describes both the physically manifested process as well as the heavenly pattern after which it was made. While the heavenly pattern was *created*, the system of seed time and harvest was *made* after its image. Things that are *created*, therefore, may be understood as existing initially in a vaporous or conceptual form, in the invisible realm, that abides at the time of a *former rain*. Things that are *made*, however, exist during a *latter rain* and grow into the full manifestation of their DNA pattern, or the pattern after which they were conceived. Things that are *made* are also understood as things that can be *shaken* as Hebrews 12:26-29 declares: "*'Yet once more I shake not only the earth, but also heaven.' Now this, 'Yet once more,' indicates the removal of those things that are being shaken, as of things that are made, that the things which cannot be shaken may remain. Therefore, since we are receiving a kingdom which cannot be shaken, let us have grace, by which we may serve God acceptably with reverence and godly fear. For our God is a consuming fire."*

Things that are shaken, therefore, refer to all those things that abide in the natural realm that are made of matter. And while many of us wait patiently in faith for matter to materialize, "*the substance of things hoped for and the evidence of things not seen*" that exists in the invisible realm has more enduring value. What can be said about faith? Our faith overcomes the world (see 1 John 5:4); faith is the substantial evidence of the conceptual seeds that exist in the heavenly realm (see Hebrews 11:1); and "*faith comes by hearing, and hearing by the Word of God*" (Romans 10:17). While these Scriptures shed light on what faith is, what it does, and how one may attain it, Hebrews 4:2 reveals perhaps the most vital truth: "*For unto us was the gospel preached, as well as unto them: but the word preached did not profit them, not being mixed with faith in them that heard it*" (KJV).

In order for the Word to bear fruit, therefore, it must be mixed with the "*substance of things hoped for,*" which actually becomes the matter of the second creation, the fulfillment of the vision or conception. This substance, as the dew that transformed into manna, is deposited and mixed with a *receptor seed* that has been sown into the heart of man by having heard and understood the Word of God. With this substance, then, and its ability to mingle with the Word, the gates of potential are opened with the result that "*all things are possible to him who believes*" (Mark 9:23). The profit alluded to in Hebrews 4:2 is the natural fruition of the initially conceived seed.

While faith is substantial, faith's potential may be considered vaporous, as the vapor phase has more intrinsic potential. Through the process of releasing heat energy, which corresponds with entropy, a vapor can transform into liquid or solid; and thereby we may acknowledge the process by which spirit transforms into flesh. This therefore indicates faith's potential, which, having *come from hearing and hearing by the word of God* is made less dense by having absorbed the Word. Should the need or desire for material substance arise, the heat energy that was absorbed—the Word—need only be released or spoken. That this kingdom of heaven system only profits those who *remain* in faith and are not shaken—which suggests that they abide in the spiritually *created* realm more than the naturally made *realm*—indicates a continual reliance on the Word and Spirit of God.

As speaking the Word of God is equivalent to releasing heat energy, hearing the Word is equivalent to absorbing it; when heat energy is absorbed, a phase of matter becomes less dense, i.e., a solid transitions to a vapor. When we hear the Word of God, therefore, we become less dense and freer and then occupy a phase with greater inherent potential.

Conversely, the less inherent potential we have signifies that we are operating outside of the realm of faith, and are instead operating according to the flesh, which signifies a bound solid phase. Having presumably achieved its state by a total *loss* of heat energy, this solid bound state in the natural parallels the state that is achieved by a total loss of the Word of God, which is, according to natural knowledge, going *with* the flow of entropy or decay: "*For when we were in the flesh, the sinful passions which were aroused by the law were at work in our members to bear fruit to death*" (Romans 7:5). When heat energy is lost, there is a corresponding loss of potential energy; in other words, that which has transformative potential dissipates because of its actualization. As heat energy can be expelled from its vaporous form and actually transition into solid matter in the process of *making* it, heat energy in the form of the Word operates according to the same principle. Only in the moment of fulfillment does the Word of God lose potential energy, as was demonstrated at the cross. Only then, when the prophecies concerning Jesus' life and death were being fulfilled, did the *Word of God* lose potential strength.

Chapter 13
Renewal

According to the process of renewal, systems erode and undergo entropy or decay, but afterward new forms of order begin to appear, such as when a man becomes born again. In their book *Order Out of Chaos*, Ilya Prigogine and Isabel Stengers spoke about this process and referred to the physical manifestations of the phase between decay and renewal as "dissipative structures." These, of course, are examples of natural phenomena that would support an evolutionist's argument that matter changes function and form over time, for which there is no argument. John 5:24, however, illustrates the heavenly process after which these dissipative structures are patterned: "*Verily, verily, I say unto you, He that heareth my word, and believeth on him that sent me, hath everlasting life, and shall not come into condemnation; but is passed from death unto life*" (KJV). Accordingly, it would appear that key evolutionary processes are, in fact, built in to the very fabric of the kingdom of God's system.

The greatest evolutionary event spoken of in the Bible is when a man becomes born again. However, how is this possible? As Nicodemus in John 3:4 asked, "*How can a man be born when he is old? Can he enter the second time into his mother's womb and be born?*" Jesus's response to Nicodemus then remains applicable even now: *"That which is born of the flesh is flesh, and that which is born of the Spirit is spirit"* (John 3:6). Becoming born again is a spiritual matter, and therefore cannot be understood by anyone operating outside of the kingdom of heaven system; it is foolishness to them. Nonetheless spiritual renewal constitutes a major pattern in the heavenly realm and is referred to in 2 Corinthians 5:17: *"Therefore if any man be in Christ, he is a new creature: old things are passed away; behold, all things are become new"* (KJV).

Being in Christ implies one thing: The state of one's being exists in a place where the Messiah resides. And if this state is considered new, one's identity is likely in a process of change, with an old identity being replaced by a new one. First Corinthians 15:49-50 mentions this passage: "*And as we have borne the image of the man of dust, we shall also bear the image of the heavenly Man. Now this I say, brethren, that flesh and blood cannot inherit the kingdom of God; nor does corruption inherit incorruption.*" As solid matter, flesh and blood decays as natural material. The kingdom of God, however, refers to the Holy Spirit, which is inherited by all those who believe that Jesus Christ is the Son of God. And with this as the criterion for becoming born again, this constitutes a vital pattern for the principle of renewal.

This principle of renewal is most evident at the most basic level, when atoms adopt a new identity. While we might say that during a process such as phase transitions, the ways in which an element appears becomes entirely new, during phase transitions, when elements change according to how they absorb or release heat, the elements' identities remain the same. During particle transformations, however, the fundamental identities are lost and are never recovered.

In the case of the elements, their identity is made known by two specific factors: The first is by the number of protons within the nucleus, and the second is by the element's spectral emission.

While the spectral emission relates to the light emitted by an atom after one of its electrons has absorbed and been stimulated by a photon of a particular frequency, the number of protons determines an element's weight as well as its intrinsic properties. The greater the number of protons, the heavier the element, and the heavier the element, the more likely it is to be radioactive and undergo radioactive decay or rather death. Paralleling the "*becoming a new creature in Christ Jesus*" pattern, however, after an element's decay, it assumes an entirely new identity.

Each element is situated along a chain of elements into which they may transform (daughter) or from which they may be transformed (mother). Where along the chain an element resides in large part is determined by the number of protons in the nucleus. As the number of protons determines an element's identity, if that number changes the element transforms into another element. However, this transformation is dependent not only on protons but on neutrons as well, for the whole system relies upon neutrons transforming into protons, or protons transforming into neutrons. However, while these composite particles are important—particularly the neutron—these changes occur at a deeper level, at the sub-particle level of quarks, specifically the up quarks and down quarks.

When a neutron, which consists of two down quarks and one up quark, transforms into a proton (two up quarks and one down), what is occurring at a deeper level is that one of the neutron's down quarks is changing its identity into an up quark. In this way, the configuration of two up quarks and one down quark provides the identity of a proton instead. And whenever the number of protons changes, so does the identity of the element. Expounding upon *identity* as a key, what becomes evident is that while the number of protons determines an element's identity, whether or not the number of electrons in that atom match the proton count determines stability.

While these processes involve up and down quarks, what is of vital value is the fact that the number of protons before or after transformation determine the identity of the element. More interesting still is the correspondence between this number of protons and the number of electrons within the atom. Please note that according to one of *LotK*'s theoretical proposals, electrons and thoughts are considered to have equivalent value and purpose. Now, if the atom is in a stable state, the number of protons and the number of electrons will be the same. If the atom is in an unstable state, these numbers will differ. With protons being able to identify an atom, and the same number of electrons able to determine stability, it would appear that electrons or thoughts would behave as stabilizing forces.

Corresponding with Proverbs 23:7 that says, "*For as [a man] thinks in his heart, so is he,*" our thoughts do tend to determine who we are. Experientially, we can also validate this. A man's thoughts about himself and the world around him determine his attitude and behavior. His thoughts will determine the words that come out of his mouth as well, and generally we know a man's identity by what he says, which is rooted in the way he thinks.

This specifically relates to the mechanics involved in having one element become a new element and by extension the actual mechanics that illustrate the transformative pattern after which God has designed a way for men to actually become "*new creatures in Christ Jesus.*" While the number of electrons (thoughts) is of vital importance to the stability of an atom (man), when the number of electrons (thoughts) differs from the number of protons, the atom (man) is vulnerable to becoming mixed with other atoms (men). The end result of this, in turn, is becoming a new compound molecule with an entirely different identity.

This mixed state of the compound molecule (or man being captured by the thoughts of others) is alluded to in James 1:8: "*A double-minded man is unstable in all his ways*" (KJV). The problem with being in this kind of mixed state is that one's identity is never able to shine, because this mixed state blurs any distinguishing qualities of one's true identity and consequent spectral emission. God has given us the ability to emit light, similar to all elements, and this light also corresponds to the resonant frequencies that are received by our *receptor seeds.* In essence, connection with God occurs most effectively when such spectral emissions are in a pure state, which corresponds with a state of single-mindedness—not closed-mindedness—having a focus on the things of God.

While an atom with an elemental identity can obtain a new identity by becoming mixed, it would appear that a pure element undergoes transformation through the process of decay. John 12:24-25 alludes to this process: "*Verily, verily, I say unto you, Except a corn of wheat fall into the ground and die, it abideth alone: but if it die, it bringeth forth much fruit. He that loveth his life shall lose it; and he that hateth his life in this world shall keep it unto life eternal"* (KJV). This Scripture describes the transformative process that occurs when a conceptual seed is planted in an invisible realm and physically manifests through the application of faith. The original seed, visible or not, no longer exists; it has transformed into something entirely new, and therefore its identity is unrecognizable; when something is no longer able to be comprehended as itself, it may, at that point, be said to have died. While this Scripture speaks about the process of planting a seed, which physically dies to itself and to its identity so that it can be provided with a new identity, both of these processes involve new birth. But the specifics involved with providing man with an opportunity to become a new creature is the precise pattern of the decay process that enables one element to transition into another.

There's an interesting correlation that can be discerned when these processes of decay occur. While the composite particles in a proton are two up quarks and one down quark, the composite particles in a neutron are two down quarks and one up quark. With the help of a mediating particle, called the W^+ boson, an up quark from a proton turns into a down quark, and the consequence is that the proton becomes a neutron. When the proton obtains this new identity, the result is a reduction of the number of protons by a count of one, and the element assumes a new identity. Similarly, if a neutron, which is made up of two down quarks and one up quark, has one of its down quarks turn into an up quark with the help of a W^- boson, the neutron turns into a proton, which increases the proton count by one. Again, a new element is born. While very similar in operation, these two decays do in fact differ, particularly in the way that energy is utilized to power the process.

The difference between these two decays then is made evident by the fact that because the neutron is larger (has more mass) than the proton, and therefore has more stored energy ($E = MC^2$), it is consequently able to make the transition by losing heat energy, which corresponds to a natural process of decay. This is an easier process, because losing energy requires no effort. When the proton transitions into a neutron, however, it requires energy from an outside source in order to make the switch.

With regards to becoming "*a new creature in Christ Jesus*," this energy is provided by the light of the Word of God, which is often initially transmitted by an electron (thought) and is able to provide the energy that enables the transformation. Isaiah 55:7-9 mentions thoughts as the elements that will enable change: *"Let the wicked forsake his way, and the unrighteous man his thoughts; let him return to the Lord, and He will have mercy on him; and to our God, for He will abundantly pardon. 'For My thoughts are not your thoughts, nor are your ways My ways,' says the Lord. 'For as the heavens are higher than the earth, so are My ways higher than your ways, and My thoughts than your thoughts.'"* Interestingly enough, the acquisition of more energy that will enable the proton to transition into a neutron occurs through the process of electron capture.

In the case of electron capture, an electron is drawn into the nucleus from the most inner electron shell—the K-shell. As a means by which a proton can become a neutron, electron capture supports the theoretical perspective that thoughts behave as electrons do, and the number of electrons and protons determine an atom's stability. As a man must change his thoughts in order to change, a proton must gather energy from an inner electron in order to make the transformation into the larger neutron. That it is an electron from the K-shell that is captured, however, leads to the assumption that the parallel thoughts that are captured by a man's conscious mind are actually already existent, but are present in a repressed form in the unconscious realm. And in actuality, this is the realm of truth, as psychologists may attest, and a domain space in which a seed as of white light, which consequently has illuminating potential, has been planted, which behaves as a man's conscious. This is also that which was spoken about in John 1:4: "*In Him was life, and the life was the light of men.*"

The light was provided to "*all*" men, and thus as a seed has conceptually, at least, been planted in every human heart. However, Paul wrote in 2 Thessalonians 3:1-2, *"Finally, brethren, pray for us, that the word of the Lord may run swiftly and be glorified, just as it is with you, and that we may be delivered from unreasonable and wicked men; for not all have faith.*" It is therefore evident that all men *do not* have faith, which leads us to assume that they are connected to the frequencies that correspond to words other than the Word of God. These words, in turn, not only do not promote faith in God, which "*comes by hearing, and hearing by the Word of God*" (Romans 10:17), but they connect these men to frequencies that are anti-God in nature. These frequencies also will recoil when in the presence of the Word or the light, as John 3:20-21 attests: "*For every one that doeth evil hateth the light, neither cometh to the light, lest his deeds should be reproved. But he that doeth truth cometh to the light, that his deeds may be made manifest, that they are wrought in God"* (KJV). The realm of truth, therefore, resides in the light of men, and the light of men is in the realm of the Word, which is the realm of Jesus, the Holy Spirit and the Father God.

While electron capture appears to represent the process whereby a man can apprehend the thoughts of his true heart, positron emission, the other way by which a proton can transform into a neutron, represents the process of expelling thoughts of an antithetical nature. As an electron's anti-particle, the positron is oppositely charged, meaning that it is repelled by what the electron is attracted to, and is attracted to what the electron is repelled by. For instance, while an electron may be attracted to thoughts that are true, honest, just, pure, lovely, and of a good report, a positron would be attracted to thoughts that are false, dishonest, unjust, ugly, and of a bad report. In essence, in order to change into a prototypical spiritual neutron, or a particle that is the key of nuclear power, one can either capture thoughts that align with Philippians 4:8, which parallels electron capture, or they can expel thoughts that oppose them.

In order for a person to change, old thoughts must be expelled and new thoughts must be acquired. What's interesting to note is that in the same way that electrons appear to serve a stabilizing purpose for the protons, when a proton transforms into a neutron a positron and an anti-neutrino are expelled, but when a neutron transforms into a proton an electron and a neutrino are emitted. Therefore, it would appear that the neutrino and anti-neutrino also behave as stabilizing forces.

Deductively, if these particles are expelled when an atom is acquiring a new identity, these particles were intimately tied to the former identity. And as electrons and positrons are anti-particles of one another, that would suggest that each serves an opposite master, which effectually controls the mind. All of this makes experiential sense; in order to change, our thoughts need to be renewed; out with the old and in with the new. Matthew 9:17 puts it this way, "*Nor do they put new wine into old wineskins.*" If a man's identity relates to his thoughts, then in order to change the thoughts must change as well. Accordingly, Ephesians 4:23-24 says, "*and be renewed in the spirit of your mind, and that you put on the new man which was created according to God, in true righteousness and holiness.*"

Chapter 14
Salted

If a man is considered stable in all of his ways, he is not, according to James 1:8, double-minded. This double-mindedness, in turn, is described for us with a picture in James 1:6; this wavering is "*like a wave of the sea driven with the wind and tossed*" (KJV). Being as a wave that is fundamentally subject to the whims of wind is another way of describing basic instability; according to particle physics, instability is described by an unequal number of protons and electrons in an atom. This inequality, then, transforms the identity of the atom into an ion.

When an atom loses or gains electrons from the valence shell, the result is that the atom becomes charged. If there are more electrons than protons it becomes negatively charged (anion). If there are less electrons than protons it becomes positively charged (cation). Under the assumption that electrons may be considered equivalent to thoughts, this condition of being ionized, having electrons given to or taken away from an atom's valence shell, signifies a person's vulnerability to being persuaded or controlled by environmental factors. For instance, when a person who is *charged* by a particular thought frequency comes into contact with another person who is also charged by another attractive thought frequency (remember positive and negative electric charges attract), a bond naturally occurs without resistance.

This lack of resistance is then directly tied to James 1:8. Someone who does not resist is as "*a wave of the sea, driven with the wind and tossed,*" and may consequently be considered double-minded. Therefore, as an element may change into another element through the process of decay, it may also be chemically transformed by mixing with another element. These bonds can be ionic with cations and anions attracted to one another, filling in or taking up electrons that make the incomplete and unsettled ions relax; or they can be covalent, where an electron or electrons are equally shared between two atoms, each fulfilling the other's lack. Second Corinthians addresses these bonds, albeit with a warning: "*Do not be unequally yoked together with unbelievers. For what fellowship has righteousness with lawlessness? And what communion has light with darkness?*" (2 Corinthians 6:14).

While a new chemical identity is not created from these bonds the way one is when atoms undergo nuclear decay, the capability of creating bonds, in general, signifies a basic reactivity. Due to a fundamental need to achieve a state of equilibrium, atoms with a valence shell of more or less than eight electrons will experience a state of imbalance. This imbalance thus accounts for an atom's instability, and like people, their natural inclination is to search until they find the missing *charge* that will fill the gap. However, this demonstrates reactivity, the opposite of which is peace. When an atom or a man is at peace, there is no hole in need of being filled. Reactivity then may be just another way of communicating instability, which is often seen in those who appear not to be *grounded* and who all too easily connect with surrounding frequencies unexpectedly. It's also interesting to note that the process of reactivity releases energy, an exothermic process, which coincides with the phase transitions that produce bound, or mixed, solids as opposed to free vapors.

While suggesting a greater bondage, reactivity also implies that electrons (or thoughts) are instrumental in sparking chemical reactions between elements, and therefore thoughts appear to be behind all volatility. This interpretation becomes particularly applicable when considering the characteristic lack of reactivity of a *pure* atom, which confirms an implication of James 1:8: If "*a double-minded man is unstable in all his ways,*" a single-minded (pure) man is stable in all of his.

Instability and incompleteness may then readily apply to anything or anyone who readily connects or binds with another. Stability and wholeness, however, signifies that there is an innate covering—as of eight electrons that stand as guards against invaders or those with whom connection in the form of a bond may be made.

The heavenly pattern with which this coincides, interestingly enough, is the account of Noah. After his obedience in building the ark and after obtaining the pairs of animals, and after the seven members of his family entered the ark, Noah, the *eighth* person, entered and closed the door. A remarkable parallel is apparent in this account. Noah was like a free electron, unconnected and vulnerable, as was his family until he joined his wife and completed the "hedge of protection."

When the four couples were bound together in a representative valence shell, with all the pairs of animals, like electrons themselves, taking up the inner shells of the ark, then and only then were they protected from the floodgates. The ark thus became a perfect symbol of a stable atom that became so because of Noah's faith in, and obedience to, the voice of God.

Patterned after an element with a potentially high atomic weight, the ark may truly be the first example of God's grace. Due to an implied high number of animals, which may be seen as an expression of electron pairs, a high number of electrons would mean a high number of protons. A high number of protons would then make the probability of a high number of neutrons greater. But if this was the case, the condition of the ark, as a representation of an atom, would probably not have been stable, as all elements above 82—which is the atomic number, or number of protons—are radioactive. However, with the proposed complete valence shell—an octet, with Noah being the eighth person—the ark *was* stable. And this is where the grace of God shines; He takes the potentially sick, unstable, reactive, and incapable and provides them with a protective shell by which they may weather the storms of life. Psalms 91 provides an apt illustration:

> *He who dwells in the secret place of the Most High shall abide under the shadow of the Almighty. I will say of the Lord, "He is my refuge and my fortress; my God, in Him I will trust." Surely He shall deliver you from the snare of the fowler and from the perilous pestilence. He shall cover you with His feathers, and under His wings you shall take refuge; His truth shall be your shield and buckler. You shall not be afraid of the terror by night, nor of the arrow that flies by day, nor of the pestilence that walks in darkness, nor of the destruction that lays waste at noonday. A thousand may fall at your side, and ten thousand at your right hand; but it shall not come near you (Psalms 91:1-7).*

Beyond the protection and grace of God evident in this illustration, it also provides an illustration of wholeness. As an incomplete atom or ion seeks to gain equilibrium by mixing with another that may supply what it lacks, this supplying of lack is what the Lord Jesus Christ came to accomplish. Matthew 9:11-12 addresses this. Answering the Pharisees who asked, "*Why eateth your Master with publicans and sinners?*" Jesus said, "*They that be whole need not a physician, but they that are sick.*" Being sick, in this context, therefore, is as an incomplete valence shell in need of another with which it may become whole.

The wholeness referred to here is the union between Jesus and mankind, with the mind of Christ (see 1 Corinthians 2:16), His thoughts, behaving as the electrons that will complete humankind's incomplete valence shell. In verse 22 of this same chapter, the subject of wholeness is mentioned again when a woman with a blood disease touched Jesus' garment. Desiring completeness, a conceptual seed was created in her mind's eyes that envisioned her whole of her disease, and she *said* within herself, "*If I may but touch his garment, I shall be whole*" (Matthew 9:21 KJV). Her thoughts joined with what she believed Jesus was capable of doing, and she touched His garment. And when Jesus turned and saw her, He said, "*Daughter, be of good comfort; thy faith hath made thee whole*" (Matthew 2:22 KJV).

Being reactive, readily joining one's thoughts with others' thoughts that are contrary to Philippians 4:8, or being sick are examples of being in a parallel condition to an unstable atom. However, joining with thoughts that correspond with the Word of God, things that do correspond with things that are true, honest, just, pure, lovely, and of a good report places one in a parallel position to a stable atom or even better yet, salt. That cations and anions are ions that have not enough or too many electrons and need each other to achieve neutrality—or rather, rest—is the chemistry at the root of Matthew 5:13: "*Ye are the salt of the earth: but if the salt have lost his savour, wherewith shall it be salted? it is thenceforth good for nothing, but to be cast out, and to be trodden under foot of men*" (KJV).

A cation and an anion, before they bind with one another, are both incomplete and require the other to attain desired neutrality; it is this union which produces the chemical composition of salt. We are therefore equivalent to an ion in need of bonding with Jesus—an oppositely charged ion who may provide us with rest, as Jesus proclaimed in Matthew 11:28-30: "*Come unto me, all ye that labour and are heavy laden, and I will give you rest. Take my yoke upon you, and learn of me; for I am meek and lowly in heart: and ye shall find rest unto your souls. For my yoke is easy, and my burden is light*" (KJV).

In essence, being provided with the label of the "salt of the earth" implies having bonded with Jesus, who, behaving as a negatively charged anion, provides us with His thoughts and thereby fills any gap. Isaiah 55:7-8 reveals God's heart concerning this union: "*Let the wicked forsake his way, and the unrighteous man his thoughts; let him return to the Lord, and He will have mercy on him; and to our God, for He will abundantly pardon. 'For My thoughts are not your thoughts, nor are your ways My ways,' says the Lord. 'For as the heavens are higher than the earth, so are My ways higher than your ways, and My thoughts than your thoughts.'*" Clearly, God's perfect will for His children is that they occupy the pattern after which a salt bond was made, joining with the mind of Christ and adopting His thoughts as their own. This is the means by which Matthew 5:13 came to pass, where the disciples or followers of Jesus become the true "salt of the earth."

Chapter 15
Raman and Rayleigh

Due to the fundamental human need to belong to others, people are naturally inclined to connect and form intimate bonds with others; however, another inclination is to resist those who don't support one's bonds. Human dynamics, which emulate God's in the heavenly realm, adhere quite perfectly to the force of electromagnetism. As discussed, man is not only actively, yet unknowingly, engaged with electromagnetic frequencies that connect with the thought seeds that have been planted in the heart, but the connections and repulsions he experiences are directly tied to the dynamics of electromagnetism. Most every interaction can be seen as occurring in response to individual electrical charges that either attract or repel one another. And when the frequencies to which man connects are considered, the charges to which people may respond not only constitute the sources of all reactivity, but also must be understood as potentially infinite.

As light is an electrical phenomenon, to which Benjamin Franklin attested, we may more fully recognize these frequencies as the root source of individual charges. Now, considering that man has been made in the image of God according to Genesis 1:26, we may reasonably assume that God, who is the light of men according to John 1:4, is electrically charged as well. So in addition to behaving as an anion, a negatively charged ion, God would manifest Himself as white light or a storehouse of excess energy. And as white light, He would encompass the entire visible spectrum, which would allow every color variant to be reflected by Him in total honesty and truth, as Hebrews 4:13 alludes: "*Neither is there any creature that is not manifest in his sight: but all things are naked and opened unto the eyes of him with whom we have to do*" (KJV).

A more important aspect of God as white light, though, is that because of the ability to reflect the truth that is in everyone, God knows and understands everyone, and this understanding breeds compassion or love. The white light of God then may be defined as the true expression of love, for it reflects the secrets within mankind; it doesn't condemn them, as Romans 8:1-2 aptly declares: "*There is therefore now no condemnation to those who are in Christ Jesus, who do not walk according to the flesh, but according to the Spirit. For the law of the Spirit of life in Christ Jesus has made me free from the law of sin and death.*" Summarily, the law of the Spirit adheres to the frequency of white light or love;

as everyone is operating according to one spectral frequency or another, God knows everyone and knows whether they are operating according to His Word and His Spirit or not. Even when people are not operating according to His frequency though, His ability to reflect their particular spectral emission breeds not judgment, but compassion and love, which defines who He is, as 1 John 4:8 proclaims: "*God is love.*"

As God is love, and we have been created in God's image, we may unequivocally validate love as intricately involved with all of the "positive potential" of which we as human beings are capable. And union with God through the mediator Jesus Christ provides the ultimate description of this potential; for one, because this union is the pattern after which salt is produced there is an implicit enablement of electrical conductance. Another potential exists in the furnishing of the light that enables mankind to fulfill God's two primary commandments: *"You shall love the Lord your God with all your heart, with all your soul, and with all your mind,"* and, *"You shall love your neighbor as yourself"* (Matthew 22:37, 39). Adhering to the ethic of reciprocity, this love may be defined as an ability to empathize with and have compassion for others, which is also the basis of the Golden Rule, "*Do unto others as you would have them do to you,*" and it assumes an active desire to alleviate the suffering of others.

This fundamental moral rule has been found in various expressions in most religions and creeds throughout the ages, thereby testifying of its universal applicability. The Golden Rule as found in the Bible, however, appears in Matthew 7:12 and says, *"Therefore, whatever you want men to do to you, do also to them, for this is the Law and the Prophets.*" Well, if "*God is love"* and has perfect love for us, and He "*so loved the world that He gave His only begotten Son, that whoever believes in Him should not perish but have everlasting life*" (John 3:16), then it stands to reason that if the Golden Rule reflects the law of reciprocity, then God wants those for whom He sent His Son to simply love Him back, and perhaps even have compassion for *Him. Ask not what God can do for you, but what you can do for God.*

In the event that Jesus was suffering, how would we even go about trying to alleviate His suffering? John 11 provides such an account when Jesus' friend, Lazarus, died:

> *Therefore, when Jesus saw her [Mary] weeping, and the Jews who came with her weeping, He groaned in the spirit and was troubled. And He said, "Where have you laid him?" They said to Him, "Lord, come and see." Jesus wept. Then the Jews said, "See how He loved him!" And some of them said, "Could not this Man, who opened the eyes of the blind, also have kept this man from dying?" Then Jesus, again groaning*

in Himself, came to the tomb. It was a cave, and a stone lay against it. Jesus said, "Take away the stone." Martha, the sister of him who was dead, said to Him, "Lord, by this time there is a stench, for he has been dead four days." Jesus said to her, "Did I not say to you that if you would believe you would see the glory of God?" (John 11:33-40)

It is clear that this event upset Jesus, but of course everyone was upset because their friend had died. But the expression of true love and compassion was not simply empathizing with their pain, it came after the stone had been rolled away and, looking up, Jesus said, *"'Father, I thank You that You have heard Me. And I know that You always hear Me, but because of the people who are standing by I said this, that they may believe that You sent Me.' Now when He had said these things, He cried with a loud voice, 'Lazarus, come forth!'"* (John 11:41-43). And bound hand and foot in grave clothes, with his face bound in a napkin, Lazarus sat up and Jesus commanded those who were around to *"loose him, and let him go"* (John 11:44). The ultimate expression of love and compassion coming from God is to "loose" those who are bound, while it would appear that what God wants from us is our trust and our faith in Him.

Similar to how we may feel with regard to our own children whom we love with a pure love, we want them to trust us and believe that we want what is best for them. We want our children to yield to our convictions, and those who have been in the position where their intentions for good were not trusted know just how painful it is. God is in that position continually. All He wants is for us to trust Him, to have faith in His intentions, and to walk with Him in resonant communion.

Our trust and faith in God, however, is contingent on our knowledge concerning what He has said. In this earthly realm, we can hardly expect to believe in anyone's intentions without first hearing words spoken with regard to a certain matter. And our estimation of their faithfulness is contingent upon having heard their words. Words become a plumb line, of sorts. Once spoken, these words draw a line from which all subsequent actions are measured. If we can superimpose the image of their actions over this line and there is no deviation, trust will follow; but if and when there are deviations, we are wary.

The fact is that faith or belief is backed by actions that correspond with spoken words. What is vitally important then is that we learn what God has said, but it is essential that we learn what He has said for ourselves. With all of the frequencies available—a.k.a. deceiving spirits to which we all are vulnerable—the need for personal connection with the one true God is absolute.

As our trust in others is often contingent upon their own motivation and commitment to align their actions with their words, God's concern for us revolves around the inclinations of our own hearts. If our motives are pure, we can be provided with God's "loving, compassionate" energy; however, if our motives are not pure, or in other words are mixed, not only may we be considered double-minded but our own energy may be accurately perceived as having been carried away, stolen, or scattered.

These processes involve transference of energy, which occurs in two fundamental ways that have been referred to as elastic scattering, or Rayleigh scattering after British physicist Lord Rayleigh, and inelastic scattering, better known as Raman scattering after Sir Chandrasekhara Venkata Raman. As it has been proclaimed that Jesus is the "*light of men*" and thus operates according to the dynamics of light, it's vital that we first learn all of the implications of this association. As a side note, as Jesus is light and light is the only element that does not have a phase state equivalent to rest, clarification to at least one scriptural reference is brought forth. In Matthew 8:20 Jesus said, *"Foxes have holes and birds of the air have nests, but the Son of Man has nowhere to lay His head."* Contrary to the age-old assumption that Jesus was poor, He wasn't referring to having no bed or house in which to rest; He was rather referring to His true identity as the essence of light.

As light, Jesus neither had a phase state equivalent to the state of rest, nor more importantly, as the Word, there was no living being available in which He might dwell, due to the disparity between energy frequencies between God and man; the Holy Spirit eventually provided this, but it was necessary for Jesus to die first before the Holy Spirit would be available to live inside believers and provide a place where Jesus, the Word, might find "rest." As a photon requires a matching energy level in an atom in order to be absorbed, so too does Jesus require a matching energy level in people's hearts in order to be absorbed and consequently have His light re-emitted through us. Having the potential energy of the Word of God and/or the Spirit of truth stored in the heart is the means by which this matching frequency is thus attained.

Light's interactions with matter are numerous. It can be absorbed, as by an electron in an atomic shell; it can be reflected, as in a mirror or in a pool of clear water; it can be refracted, bending off in another direction as it enters a different medium than the one from which it was sent; or it can be affected by interfering waves. All of these interactions ultimately affect the light that emanates from the object/person after interaction with the light. Substituting the Word of God/Jesus for what we know as light in all these contexts allows us to see more clearly the power of the Word. Similar to the interference of water waves, when light waves collide the destructive or constructive interference changes the phase speed of the waves or resulting wave.

Unlike water waves, light's frequencies may be determined by the colors they emit, with red signifying a frequency much slower than blue, for example.

By extension, we ourselves emit light; as an element is known by its spectral emission, so are we, and our emissions are dependent upon the frequencies that we admit into our heart. Therefore, the light which is emitted shows the condition of our hearts, which is how we are known by God. This is how "*the Lord knows those who are His*" (2 Timothy 2:19; see also Acts 1:24), by the color of our spectral emission.

Effectually then, under the condition that we absorb a light that is emitted in our direction, it has the opportunity to be absorbed and excite electrons (thoughts) in the process with an inevitable re-emitting of light of the same frequency (reflection); herein lies the root source of connection. When we refract a light, however, there's nothing in us that receives of that light and it simply bounces off in another direction.

Similar to how a ball is spun off in another direction when hit by a cue ball in the game of eight ball, the process of refraction sends light off in another direction when it hits an object. Describing a quick surface connection and a subsequent deflection, the concept of Rayleigh scattering or elastic scattering is commonly referenced in the Bible with the mention of sheep being scattered. When the shepherd isn't tending to the sheep's needs of being led to green pastures, the sheep are placed in a position of taking in whatever will provide them with energy. Consequently, they are easily scattered and are vulnerable to following other men. When this analogy is used, however, it is specifically referring to pastors and teachers. Those sheep who have been led to feed on the green pastures of the Word of God have invariably ingested the Word, and having the Word of God within their hearts they will only obey the true shepherd. Those who teach any other gospel, however, will be accounted as imposters who lead the sheep astray.

There are many who sound and look like shepherds, but they don't do what only a true shepherd is able to do, that is, provide His sheep with meat (the Word of God) that teaches them how to overcome the world with faith and how to live a life pleasing to God. Being fed with the Word of God is the only way that will enable the sheep of God's pasture to grow strong. This strength, in turn, will be apparent when the sheep resonate only with the Word of God, which will ultimately direct their paths. Jeremiah 23:1-2 responds to shepherds who fail to lead their sheep: "*'Woe to the shepherds who destroy and scatter the sheep of My pasture!' says the Lord. Therefore thus says the Lord God of Israel against the shepherds who feed My people: 'You have scattered My flock, driven them away, and not attended to them. Behold, I will attend to you for the evil of your doings,' says the Lord.*"

Matthew 9:35-36 alludes to this scattering as well, but shares with us also the heart and doings of the real shepherd, Jesus: "*Then Jesus went about all the cities and villages, teaching in their synagogues, preaching the gospel of the kingdom, and healing every sickness and every disease among the people. But when He saw the multitudes, He was moved with compassion for them, because they were weary and scattered, like sheep having no shepherd."*

While God is love, love is equivalent to white light, and the Father, the Word/Son, and the Holy Spirit are One. What becomes evident is that when the light of the true Word of God is not emitted, the hungry will invariably absorb whatever light is sent their way. As a consequence of this absorption, however, not only will any re-emitted light reflect this impure light, but when the light of the true Word of God is emitted, the previously absorbed light with interfere with any potential absorption of the light of God's Word. With nothing of the true Word of God or God's truth within the heart, there will be no consequent resonance with it. Deceptive seeds, in the form of other words and/or intents, are instead absorbed and become receptors, which causes resonance with similar deceptive frequencies. When the Word of God is sent forth, these deceptive seeds will likely respond with repulsion, and will effectively cause a scattering from the Word to occur. This is what occurs with bad or incorrect teaching; deceptive seeds are planted and seek to resonate with other similar frequencies, and if there is not a turning, or repentant response, the scattered can become lost.

The grounds (hearts) of people who have heard and internalized these (impure) words, therefore, become transparent to the true Word of God, having no receptor within them that can accept or "hear" the Word of God, which has as its witness the Spirit of truth. Thus, they become vulnerable to scattering as is mentioned in Ephesians 4:14: *"that we should no longer be children, tossed to and fro and carried about with every wind [wave] of doctrine, by the trickery of men, in the cunning craftiness of deceitful plotting."* Those who are ill prepared by having heard false doctrine and/or testimonies, or those who in their defensiveness mock the Word, discounting its validity, are equally able to be scattered or led astray, for both constitute enemies of the truth: *"For behold, Your enemies, O Lord, for behold, Your enemies shall perish; all the workers of iniquity shall be scattered"* (Psalms 92:9).

The usual mention of scattering in the Bible describes a process which appears to be the heavenly pattern for elastic scattering or Rayleigh scattering. While mentioned in the context of sheep being scattered, which provides a heavenly parallel pattern, the process of Rayleigh scattering entails light being scattered by individual atoms or molecules. This occurs most readily when light passes through gases, but also occurs when light passes through transparent liquids or solids. Rayleigh scattering occurs as a result of an oscillating light wave as it acts on a particle's inner charges, which respond with attraction or repulsion to the light. This causes polarization of the charges within the particle and it becomes a small radiating dipole whose radiation is seen as scattered light.

If an incoming light source causes polarization, we may assume that the particle is "*as a wave of the sea driven with the wind and tossed*" (James 1:6), as it is the charges *within* the particle that ultimately cause it to polarize. As a response that entails the bound charges within an atom—nucleons and electrons—to become separated, this polarization can be understood as the physics behind confusion or double-mindedness. As an electric dipole moment, this separation between charges clearly indicates an identity crisis of sorts, particularly with consideration that the number of electrons in a *stable* atom matches the number of protons; if an atom has charge it has more or less electrons than protons. Consequently, if there is a separation between these particles, identity or at least clarity of purpose would appear to suffer. Having been patterned according to a heavenly spiritual truth, this separation may thus be understood as a separation between the heart of a man and his thoughts, the consequence of which opens the door to frequencies that lead to double-mindedness, confusion, and/or disobedience. And while particles that are tossed by any of these "light waves" become small radiating dipoles whose light is seen as scattered, a similar scattering of thoughts is apparent in the confused.

The other kind of scattering is inelastic scattering or Raman scattering, which instead of describing how light can be carried off in a seemingly directionless manner, describes the transference of internal energy to and from a source of light. Discovered as an effect that occurs when incident photons collide with matter, Raman scattering relates to phenomena where the photons' energy is either increased or decreased after contact with matter. After collision, if inelastic scattering has occurred, either the photon's energy will have transferred to the matter or the matter's energy will have transferred to the photon. In the case of matter having gained the photon's energy, the loss of energy in the photon makes its wavelength longer and its frequency slower (red shift). However, when matter's energy is transferred to the photon, the light's wavelength becomes shorter and the photon's energy increases (blue shift). From the perspective of an observer, red shift occurs whenever a source of light travels further away, whereas blue shift occurs when a light-emitting object moves toward the observer. Consequently, the transferring of energy from photon to matter or matter to photon has made Raman scattering a significant tool in determining distance of planetary bodies as well as the composition of solids, liquids, and gases.

The relevance of both Raman and Rayleigh scattering comes to full fruition as we consider what the Bible says about scattering. While most biblical references allude to the process of Rayleigh scattering, Proverbs 11:24 provides an example that parallels Raman scattering: *"There is one who scatters, yet increases more; and there is one who withholds more than is right, but it leads to poverty."* This Scripture illustrates two different aspects of inelastic scattering: The first part alludes to the transfer of an incident photon's momentum, i.e., *"one who scatters,"* which may also be seen as an effect of elastic scattering. The next statement, however, "*yet increases more*" can be understood in terms of what occurs when a photon transfers its energy to a particle of matter. The matter's energy increases even though it has also been displaced by the light's momentum. When we come into contact with colliding frequencies (spirits), which most often operate through human bodies, we either gain light energy, lose it, or achieve a state of neutralization, which is most readily experienced as peace and harmony.

We all emit a light, whether light or darkness as Matthew 6:22-23 suggests: "*The lamp of the body is the eye. If therefore your eye is good, your whole body will be full of light. But if your eye is bad, your whole body will be full of darkness. If therefore the light that is in you is darkness, how great is that darkness!"* The quality and character of the light that we emit, therefore, reveals the condition of our hearts. And while a scattered light would point to an uncommitted and double-minded condition, a purposeful beam of reflective light, as close to pure white as possible, would reveal a single-minded and committed heart. A molecule's internal structure, therefore, is made known by how its bound components respond to an external field, and the fundamental probability of a person being scattered is connected to the internal condition of their heart, which is affected by the frequencies that he or she admits into their lives.

In terms of the Word of God and inelastic scattering, the Word never actually loses energy; it is equal to light in every way. And the one property that light shares with no other phenomenon is that light has no rest mass; it travels at one speed—the speed of light—and cannot achieve a state of physical rest. As Jesus is considered not only the Word of God but also the light of men specifically according to John 1:4, Jesus Himself has zero rest mass.

Jesus, therefore, like light itself, is never truly capable of losing energy; He is, however, susceptible to being redirected elsewhere, where He, His Word, will more readily be absorbed. The light of the Word of God experiences a similar displacement. Having been created in God's image in earthly vessels, people therefore are simply objects into and out of which light energy flows. In the same way that an atom either loses or gains a photon's energy during the process of inelastic scattering, so too do we either gain or lose the light energy of the Word of God.

With consideration of human imperfections, the "*light of men*" (John 1:4), which is the Word, is refracted to a certain extent when it is emitted by a human being. Due to the fact that we are imperfectly "solid," our absorption of the light is limited. With this consideration, we may more fully understand this principle of inelastic scattering. As mere mediators of the power of the Word, those who preach or teach the Word will feel the inelastic scattering effects. While the power of the Word is disseminated, a corresponding draining of energy occurs within a minister's own body that will often only be realized later. However, to those who actually *hear* the Word, while they are moved by the momentum of the incoming light—just as a particle involved in Rayleigh scattering would be when struck by a photon—they also gain a portion of the light's energy. This provides an example of "*one who scatters, yet increases more*" (Proverbs 11:24).

When someone accepts Jesus into their lives, His light energy fills their bodies and they gain His strength, which is substantiated in Philippians 4:13: "*I can do all things through Christ who strengthens me*." As a photon's energy is given to a particle and experiences a loss of energy, so too do those who emit the light (Word) to others experience an energy loss. This loss, however, becomes someone else's potential gain as some will hear and become a storehouse of the light's energy (Word) themselves. Luke 8:18 makes mention of this process: "*Therefore take heed how you hear. For whoever has, to him more will be given; and whoever does not have, even what he seems to have will be taken from him.*" The more Word someone has in his/her heart, the more resonance will occur when the Word of God is heard. And as "*faith comes by hearing, and hearing by the Word of God*" according to Romans 10:17, as hearing increases, the light energy of faith begins to multiply. However, when there is no Word within the heart, there is consequently no resonance when the Word of God is heard, and therefore any light of faith that might have been gained is lost instead, and the person is as one who is scattered.

Researchers have discerned that planetary bodies are moving away from us due to a corresponding red shift of the traveling light seen from our distance. The lengthening of light's wavelength, from the blue end to the red end of the spectrum, indicates the furthering of position from an observer. Inelastic scattering has thus provided researchers with a tool to discern celestial inclinations; perhaps it may one day be a tool that may discern the inclinations of individual hearts as well.

Chapter 16
Transference

"A sower went out to sow his seed. And as he sowed, some fell by the wayside; and it was trampled down, and the birds of the air devoured it. Some fell on rock; and as soon as it sprang up, it withered away because it lacked moisture. And some fell among thorns, and the thorns sprang up with it and choked it. But others fell on good ground, sprang up, and yielded a crop a hundredfold." When He had said these things He cried, "He who has ears to hear, let him hear!" Then His disciples asked Him, saying, "What does this parable mean?" (Luke 8:5-9).

"Now the parable is this: The seed is the word of God. Those by the wayside are the ones who hear; then the devil comes and takes away the word out of their hearts, lest they should believe and be saved. But the ones on the rock are those who, when they hear, receive the word with joy; and these have no root, who believe for a while and in time of temptation fall away. Now the ones that fell among thorns are those who, when they have heard, go out and are choked with cares, riches, and pleasures of life, and bring no fruit to maturity. But the ones that fell on the good ground are those who, having heard the word with a noble and good heart, keep it and bear fruit with patience. No one, when he has lit a lamp, covers it with a vessel or puts it under a bed, but sets it on a lampstand, that those who enter may see the light. For nothing is secret that will not be revealed, nor anything hidden that will not be known and come to light. Therefore take heed how you hear. For whoever has, to him more will be given; and whoever does not have, even what he seems to have will be taken from him (Luke 8:11-18).

The scenarios spoken about in Luke 8 pertain to absorbing the Word and Spirit of God fully, partially, or not at all, with full absorption belonging to those who are "*rooted and grounded in love*" (Ephesians 3:17) as God is love. The results of these levels of absorption of the Word (the "*light of men*" in John 1:4) correspond to the ability to gain the substance of faith—a connection substantiated in both Romans 10:17 and Hebrews 11:1.

The level of absorption of faithful substance, which comes by hearing the Word, thus determines how fruit, or the manifestation of what is hoped for, is multiplied in one's life. *"The kingdom of heaven is like a mustard seed, which a man took and sowed in his field, which indeed is the least of all the seeds; but when it is grown it is greater than the herbs and becomes a tree, so that the birds of the air come and nest in its branches"* (Matthew 13:31-32).

The Word of God is referred to as the seed in the parable of the sower, and it is planted in differently conditioned soil. Therefore what becomes evident is the importance of the ground on which the seed is sown, whether it has been prepared or not. The ground, in this context, is analogous to the heart; and the heart, as the central part of a person, governs what words are listened to, what words are ignored, what words are acted upon, and what words are believed as true and ultimately retained. Preparing the heart, therefore, involves opening one's mind to the frequencies that correspond to the Word of God and asking for understanding; as Matthew 7:7-8 says, *"Ask, and it will be given to you; seek, and you will find; knock, and it will be opened to you. For everyone who asks receives, and he who seeks finds, and to him who knocks it will be opened."*

A variation of Luke 8:18 is written in Mark 4:24: *"Take heed what you hear. With the same measure you use, it will be measured to you; and to you who hear, more will be given."* Therefore the words that are *heard*, esteemed as valid, interpreted as correct, and therefore believed as true will determine not only the path upon which one travels, but also all future knowledge and understanding that will be gained. Consequently, anything that deviates from the truth that is spoken of in the Bible, if it is accounted as having more value than the Word of God at any time, will simply swallow up any receptor seed of faith that had been sown in the heart, as well as any subsequent understanding.

As light energy is gained by matter when it collides with a photon, so is faith gained by an interaction with the light of the Word of God when it is spoken and subsequently heard by a believing heart. This constitutes good ground. However, it is also true that the energy in matter can be transferred to a photon when a collision occurs[5], which alludes to the process of faith being stolen or stripped of its substance. The matter, or person, is consequently left drained while the incident photon or frequency/spirit gains that energy. This alludes to the wayside seed—the Word that was stolen by the Devil. The seed sown on rock and on thorns more specifically alludes to the local magnetic forces that draw us toward them and away from the one true magnet—God.

In the event, however, that matter gains the pure white light energy of love, by which faith operates (see Galatians 5:6), then that person (matter) may collide with others and transfer that light of love to others. Never truly decreasing, light energy adheres to the law of conservation and is only capable of being displaced

5 "Inelastic Scattering," IUPAC Gold Book, Last modified August 19, 2012, accessed January 28, 2014, http://goldbook.iupac.org/I03025.html.

for a time before returning to a state of equilibrium. First Corinthians speaks about this quality of the white light energy of love when it says, *"love never fails"* (1 Corinthians 13:8). As God is love, hearing and believing the Word of God connects one to the frequency of white light.

Chapter 17
Polarities

"*In the beginning God created the heavens and the earth. The earth was without form, and void; and darkness was on the face of the deep. And the Spirit of God was hovering over the face of the waters. Then God said, 'Let there be light'; and there was light"* (Genesis 1:1-3). From this moment forward, the experience of polar forces on earth became evident: Light broke through the darkness; form devalued the void; light was esteemed as suitable for the desired end, or in other words was seen as *good*; and darkness, lacking creative potential, served no desirable end and was separated from the good. Alluding to a vision of what God wanted to have come to pass, this first chapter of Genesis is an account of God calling those things that were not as though they already were (see Romans 4:17); it is therefore an account of the seeding of a conception, as by those who *let*, or as by those who "*write the vision, and make it plain upon tables, that he may run that readeth it*" (Habakkuk 2:2 KJV).

As the power of light broke through the darkness and was consequently separated from it, this separation initialized the experience of polar extremes, where two fundamentally opposed systems of government began to become apparent: The carnal system, which is governed by Satan and his angels, i.e., frequencies that do not glorify God that pervade the atmosphere; and the kingdom of heaven's system, which travels on a frequency of pure white light (love), perpendicular to the surface of the earth as in a 90 degree angle in perfect righteousness. This system operates on a foundational premise that it is by God's gift of grace that we are saved through faith (see Ephesians 2:8), and it is by faith that we can be accounted as righteous, as Hebrews 11:7 says, "*Noah...became heir of the righteousness which is according to faith*" because he believed God and obeyed His command.

The domain of light, therefore, operates according to the principles of grace, faith, righteousness, and obedience, while the domain of darkness functions by factors within the parameters of interference, fear, unrighteousness, and disobedience. On one level, while the kingdom of heaven operates according to the premise that "*faith works by love,*" the world's carnal system functions according to fleshly lusts and fearful disobedience. The fear of not obtaining what others may have, of losing what one has, or of taking responsibility for one's actions renders one vulnerable to a disobedient heart—disobedient to operating according to love, that is. When we are afraid of reaping what we have sown or losing what we have gained or not gaining what we see others have, we switch to a defense mode that operates outside of love, where our own needs and desires become the only focus.

When we understand God's love for us though and are fully persuaded to operate according to the fruit of the Spirit, we joyfully obey the governor of the kingdom of heaven's system—Jesus, the Word of God—knowing that His system will prevail. While the kingdom of God, which is the Holy Spirit—*"The kingdom of God is within you*" (Luke 17:21)—finds expression through the fruit of the Spirit, which is "*love, joy, peace, longsuffering, kindness, goodness, faithfulness, gentleness, self-control"* (Galatians 5:22-23), the kingdom of heaven defines God's system of government.

Revealed through parables, the kingdom of heaven is "*like a man who sowed good seed in his field, but while men slept, his enemy came and sowed tares among the wheat"* (Matthew 13:24-25); it is like a man who went out hour after hour and hired laborers to work in his field and gave them all the same reward; it is like a king who was making a marriage for his son and sent his servants to invite guests but they all made excuses as to why they couldn't make it, so he invited others instead so that his son's marriage would be furnished with guests. And it's like ten virgins who went out to meet the groom, but only five were prepared with oil in their lamps. And when asked for their oil, the five who had brought oil told those who had none to go and buy it for themselves, but when they had gone, the groom came, and the door to the marriage was shut. And only those who had been prepared were let in.

The kingdom of heaven's system thus describes an overall framework that is governed by light trying to recover those who have been captured by a system of darkness, and it describes a need to have one's heart prepared to respond to the governor of this system—Jesus. While a seed of light was planted in all men in the form of a conscience—"*In Him was life, and the life was the light of men"* (John 1:4)—without that seed being watered by the Word of God or by God's Spirit, that seed can die. And if that seed dies through hearing faithless words, that light becomes darkness. Instead of growing in reflectivity, toward pure white, the *light* becomes absorbent, like the darkness of a black hole.

Referring to this light, Matthew 6:23 says: *"But if your eye is bad, your whole body will be full of darkness. If therefore the light that is in you is darkness, how great is that darkness!*" The polar extremes of light and dark can then be understood in the context of the polar extremes of reflection and absorption. As love and white light reflects every other spectral emission, hatred, fear, and black light absorbs or swallows every other spectral emission and drains the entities of their light force. The kingdom of God (Holy Spirit), therefore, is that which has been provided us, so that we *"may proclaim the praises of Him who called you out of darkness into His marvelous light"* (1 Peter 2:9).

There are, therefore, two fundamental polar extremes: The darkness that was evident upon the face of the deep in Genesis, and the light, which wasn't understood by the darkness. Interestingly, this lack of understanding alludes to the probable fear that darkness had of light. As a cockroach runs from the light in horror, scrambling for the shadows, anything that agrees with darkness will invariably resist and fight against anything that agrees with light. An account of a meeting between Jesus and two men who were possessed with devils as they came out of the tombs is revealed in Matthew 8:29: "*And, behold, they cried out, saying, What have we to do with thee, Jesus, thou Son of God? art thou come hither to torment us before the time?"* (KJV). Darkness has a healthy fear of the light because it knows that the white light of love, particularly, will expose every secret that is hidden in the heart.

While the polar opposite domains of light and darkness separate the kingdom of heaven's system from the world's carnal system, as electromagnetic bodies, people are caught in the middle, magnetically drawn by both. And while "*we shall all stand before the judgment seat of Christ"* (Romans 14:10), according to 2 Timothy 2:20, "*in a great house there are not only vessels of gold and silver, but also of wood and clay, some for honor and some for dishonor."* Therefore, the words that we listen to and believe as true will fundamentally determine to which domain we will gravitate. Jesus' declaration of what the parable of the sower meant reveals this overall framework most concisely: "*He who sows the good seed [those who honor Him] is the Son of Man [Jesus]. The field is the world, the good seeds are the sons of the kingdom, but the tares are the sons of the wicked one [those who dishonor Him]"* (Matthew 13:37-38).

Therefore, both those with receptor seeds for the Word of God, in which is life, as well as those who have receptor seeds for dark frequencies have been planted in the earth. So it comes down to an initial choice of which *light* frequency we will be connected to—either white light or dark light, as Deuteronomy 30:19 alludes: *"I call heaven and earth to record this day against you, that I have set before you life and death, blessing and cursing: therefore choose life [white light], that both thou and thy seed may live"* (KJV).

God's favor and blessing is on those who resonate with His white light frequency; having tasted of His goodness, they expectantly wait for His return. They have "*the substance of things hoped for, the evidence of things not seen.*" They hold the conceptual seed of what is hoped for, a potential vapor, as if it were already substantially solid. And while this defines faith, there would be no foundational structure without hope. Like an armature, a frame upon which clay is molded, hope provides a foundational layer: "*And hope maketh not ashamed;*

because the love of God is shed abroad in our hearts by the Holy Ghost which is given unto us" (Romans 5:5 KJV). The kingdom of heaven, therefore, operates according to faith, hope, and love, with hope being a measure of expectancy. However, because of the world's system and its ties to darkness, instead of any initial hopeful renderings transitioning into full-blown faith, they have the potential of being rendered ineffectual by transitioning into a state of doubt and unbelief.

In essence, the Word of God generates hope that leads to faith, and all words that contradict God's Word degenerate hope and annihilate the opportunity for faith to grow. An end result of this degeneration of hope is a redirecting of the substance of potential faith. This redirection, in turn, commensurate to elastic scattering, propels us in the polar opposite direction toward unbelief and fear, which according to Revelation 21:8 are considered by God equally unholy: "*But the cowardly, unbelieving, abominable, murderers, sexually immoral, sorcerers, idolaters, and all liars shall have their part in the lake which burns with fire and brimstone, which is the second death."*

What becomes fundamental to understand is that the generation of doubt and unbelief originates in darkness. In this domain exists the antichrist spirit that wars against the spirit of truth, and one of the primary goals of the antichrist spirit is to quench any means by which people might overcome the gravitational pull of doubt, unbelief, and death. According to 1 John 5:4-5, *"For whatever is born of God overcomes the world. And this is the victory that has overcome the world—our faith. Who is he who overcomes the world, but he who believes that Jesus is the Son of God?"*

Through these Scriptures we are made aware of the fundamental battle that rages. If the victory of overcoming comes by faith in the fact that Jesus Christ is the Son of God, we know that anything that tries to interfere with the acquisition of this truth has, as its primary motive, the need to block this knowledge from taking root in our hearts. The antichrist spirit wants victory instead, and the only way to achieve it is by making sure that the seed of faith doesn't grow; consequently, Satan works on the ground of our hearts and makes every attempt to ensure that the ground of our hearts doesn't have opportunity to become *good.*

In reference to Luke 8:5-9, it appears that the means of infiltration into our hearts is through temptation as well as cares, riches, and the pleasures of life. If we can be deceived and manipulated into believing that true happiness comes from the objects and activities of this earth, we can be swayed into placing our hope in *them*. And in line with the first law of thermodynamics, where rather than dissipating an amount of energy taken from one place simply is displaced somewhere else, the consequence of all hope residing in the things of this earth results in no remaining hope. Warring against the spirit of truth, the antichrist spirit lies and deceives, as was seen in the garden of Eden. All of these earthly temptations and cares, in effect, become like magnets that draw us toward themselves.

When our hearts are not grounded in the Word of God or corresponding frequencies, we invariably become susceptible to their pull. Consequently, our acceptance of certain words or messages as true (those that we *believe* as true) determine the frequencies, or spirits, that we allow to govern our lives. Accordingly, the principles of both elastic as well as inelastic scattering apply, but the mechanistic force of magnetism is applicable as well. We can be drawn toward the things of this world, or we can be drawn toward God; we can resist the pull of the world, or we can resist the pull of God. This principle of magnetism in particular and electromagnetism in general is thus the force patterned after the invisible force that determines all human inclination.

Having been patterned after a heavenly force, the earthly force of magnetism thus emulates the pushes and pulls that occur in the heavens. Following this assumption, God Himself assumes the identity of the one true *global* magnet and can be actually felt with the wearing of a magnetic belt. As a conceptual anion whose electron charge outweighs any positive charge, God can join with those who desire to be the salt of the earth, and He guides those who are more electronegative, or have an affinity for electrons; it is consequently this force with which He draws us toward Himself. Having provided us with free will, however, the choice is ultimately ours as to whether we will obey His magnetic draw or whether we will reject it. Therefore, while God seeks to draw us toward Himself, we are simultaneously being drawn by other magnetic forces that operate on a local level.

Our ability to be pulled by either God or any of the local magnetic forces is contingent upon the receptor seeds that have been sown in one's heart. While receptor seeds in one way operate as electrons in an atom that will respond to a light of a similar frequency, characterized as having charge, they will also attract, be attracted to, repel and be repulsed by other charges that are likewise interacting with light frequencies. In addition, if an atom is considered stable and complete, it will have no unpaired electrons (thoughts); when atoms are incomplete, however, they do, and they readily seek to form a bond with others through electromagnetic attraction in order to become complete. When an atom is charged, however, it implies that it has unpaired or free electrons, which means that it is incomplete and unstable. Therefore anything magnetically susceptible is suspect of being potentially reactive and therefore readily capable of connecting to any of the local magnetic forces. God, however, as the one true global magnet, provides the magnetically attracted with a stabilized, consistent, and fully grounded light frequency with which to bind and with which they may fulfill their true potential and become the salt of the earth.

As the electromagnetic force comprises both electrical and magnetic dynamics, both the electric and the magnetic fields respond to the activity of the other. For instance, when the electric field changes, a corresponding change within the magnetic field occurs, and when a change in the magnetic field occurs, the result is a corresponding change in the electric field. Considering that electrons appear equivalent in value and function to thoughts, we can validate this fact by realizing that when our thoughts about a subject change, the feelings of like or dislike about the subject change as well. The same applies to those times when our feelings about something change; our thoughts about it also change. Therefore, our thoughts function in the domain space of an electric field, whereas our feelings function in the domain space of a magnetic field. In summary, the electromagnetic field is governed by light that affects human thought and feeling, which emulates the kingdom of God's way of leading, guiding, and correcting those who adhere to the kingdom of heaven's system.

The outstretched hand of God doesn't beat or ridicule, but rather brings out of bondage and brings to lands flowing with milk and honey. His is a gentle hand from which the force of electromagnetism emanates and causes the obedient to heed the electromagnetic pushes and pulls and attractions and repulsions, and leaves the disobedient vulnerable to the local magnetic fields' persuasion, with the potential of being pushed and pulled, this way and that way, *"like a wave of the sea driven with the wind and tossed"* (James 1:6 KJV).

Chapter 18
Spin Waves

"Which of you by worrying can add one cubit to his stature? So why do you worry about clothing? Consider the lilies of the field, how they grow: they neither toil nor spin" (Matthew 6:27-28). This explicit referral to the labor-intensive process of making fiber out of grain undercuts a spinning that is perhaps more relevant. Considered a *degree of freedom,* this spinning, along with momentum and position, defines how a particle *may* occupy and travel in space. So the questions of where one is and where one is going are implied, suggesting potential intentionality, or choice.

However, another reference to spin refers to the particle's internal spin direction, with either right-handed or left-handed spin options available. And while right-handed particles spin in the same direction as the field in which the particles are traveling, left-handed particles do the opposite. Instead of being caught up in the dominant field's persuasion, these particles resist the dominant field's direction and spin against the current.

Interestingly enough, with reference to particle conversion—the process where neutrons and protons transform into one another, which occurs during the weak interaction—the only particles that have been documented as experiencing this conversion are left-handed particles. This process, of course, parallels the heavenly pattern of how people may "*become new creatures in Christ Jesus.*" It is evident, therefore, that something resides at the core of these left-handed particles that is resistant to the dominant flow, and this may readily apply to those who prepare their hearts and turn from cultural persuasion. Instead of being led by the carnal world system, they instead are led by an internal resistance or prompting, which allows them to operate according to the kingdom of heaven system. Accordingly, John 3:6-8 provides an illustration: *"That which is born of the flesh is flesh, and that which is born of the Spirit is spirit. Do not marvel that I said to you, 'You must be born again.' The wind blows where it wishes, and you hear the sound of it, but cannot tell where it comes from and where it goes. So is everyone who is born of the Spirit."*

Describing one parameter of a particle's quantum state, the spin of a particle not only determines it tendency to follow or resist the dominant flow, but it also generates spin waves that are considered disturbances in magnetic material. As the ordering of spin states is vital for a metal to become magnetized by generating heat by way of motion, spin waves reduce the ordering of these states and thereby a permanent magnet's magnetization. While analogous to the vibrational modes caused by phonons, spin waves are governed by both internal and external fields that cause changes in rotational movement around an axis. This suggests that "spinning" is a product, not only of environmental fields, but also of an internal condition of the particle, which appears to have acquired this internal energy in collisions with other particles, or in other words through an inelastic scattering process. This in turn would suggest that another particle's energy frequency has possessed the particle and provided it with the evident excitation. Interestingly enough, one such particle that has an effect on spin waves is the neutron, which can both create as well as annihilate them. This finding, of course, supports the neutron's neutrality, because its effects are contingent on another apparent variable.

As spin is considered a degree of freedom, if this concept were to be extended to apply to human beings, the notion of choice would come into play. And if this were the case, the disturbances that a particle/person caused would be directly connected to the choices made in terms of their position, momentum, and direction of spin, or chirality. While position is an inherently relative measurement, with regards to God's kingdom one's position is most adequately measured by the heart's intentions, its motives, and whether they are pure or mixed. Momentum, on the other hand, is related to Isaac Newton's third law—the law of conservation of momentum, which defines momentum as being the product of mass and velocity and implies that momentum has a direction or path that is followed.

The combination of both momentum and the extended interpretation for position might lead to an applicable conclusion: Both intentionality and purpose may definitively define our own degrees of freedom—where we are (position), what is motivating us (momentum), and whether we are going with or against the dominant flow (spin). All of these would appear to have an effect on the level of attraction to the global magnet—God—with a greater generation of spin waves corresponding to a greater distance from God. This also would imply that the closer one is to God, the less turmoil is experienced, which corresponds with John 14:27: "*Peace I leave with you, My peace I give to you; not as the world gives do I give to you. Let not your heart be troubled, neither let it be afraid.*"

Being able to be both created and destroyed by neutrons suggests that along with being the key to nuclear power, neutrons are neutral vessels that appear to have power similar to the power of words. As words can bring peace and joy and be an expression of the love of God and mankind, thereby bringing peace, they can also destroy peace by wreaking havoc on our inward thoughts. Similarly, neutrons appear to be able to destroy the disturbances, bringing restfulness, as well as create the disturbing spin waves that cause a state of unrest. As words penetrate through to the heart, so too do neutrons, as their neutrality slips on through and past any charged source; without a built-in charge, they neither attract nor repel other charges and are thus free to interact with the nucleus of an atom, or by extension the *heart* of a man.

When neutrons are absorbed, they penetrate beyond the surface, cause that which absorbed them to split (fission), and are involved in a process of self-reproduction whereby more neutrons are created and consequently expelled and scattered. The function of neutrons, in this sense, appears to correlate precisely with that of words. An interesting parallel between the mean lifetime of a free neutron and the average absorption of words by man is also apparent; the mean lifetime of a free neutron is 881.5 +/- seconds, or about 14 minutes and 42 seconds, while the average conscious attention span for a person listening to words is at best about the same. Nonetheless, free neutrons, like words, are easily scattered and ultimately seek absorption. Therefore, if neutrons were to be considered particles that actually transmit audible as well as inaudible vibrations, as virtual thought carriers their effect on creating and annihilating spin waves would make sense: Our thoughts are readily recognized as the source of our stress as well as our well-being; they have a demonstrable effect on the physiology of the human body. We can think ourselves happy or we can think ourselves sick.

Considered to be a result of thermal excitation, the presence of spin waves implies that inelastic scattering has occurred and energy has been captured from a potentially unstable source. With true stability adhering to the pattern of a salt bond—which is patterned after an intimate relationship with Jesus, the Word of God, whereby it becomes possible to become the salt of the earth—one thing becomes evident: Those who are easily scattered likely have little Word within themselves that would otherwise act as a stabilizing force. As an electron appears to stabilize a neutron and a positron appears to stabilize a proton, the Word of God appears to stabilize one who is in need of a donation of thoughts by which their minds may become renewed.

In confirmation of this renewal occurring in response to Jesus, the Word of God, Ephesians 4:21-24 says, "*the truth is in Jesus: that you put off, concerning your former conduct, the old man which grows corrupt according to the deceitful lusts, and be renewed in the spirit of your mind, and that you put on the new man which was created according to God, in true righteousness and holiness.*" And Romans 12:2 adds, "*And do not be conformed to this world, but be transformed by the renewing of your mind, that you may prove what is that good and acceptable and perfect will of God.*" With the will of God being made known by the Word of God, Jesus, it is evident that the renewal of one's thoughts as well as the words routinely spoken comes from an intimate bond with Jesus, the Word of God. Hereby, it becomes possible to truly "*have the mind of Christ*" (1 Corinthians 2:16).

When thoughts and conversations are not renewed, however, the consequent lack of saltiness, which characterizes a true connection with the Word of God, establishes vulnerability to being overcome, and this occurs through the avenue of thoughts and words. It's important to realize that there are many "scatterers" whose mission is to impede any bond with the Word of God. Ephesians 4:28 addresses this: "*Let him who stole steal no longer.... Let no corrupt word proceed out of your mouth, but what is good for necessary edification, that it may impart grace to the hearers.*" From this Scripture, what becomes apparent is that words have as strong a capturing effect as someone coming and physically stealing our property. It's important, therefore, to become aware of one's own words and to take heed to what is listened to and regarded as true, for words can destroy both us as well as others; in the same way that an incident photon can steal energy from a particle of matter during inelastic scattering, words can steal our peace, our joy, and our faith, rendering the promises of God of no effect.

Regarding the ability to ensure that the promises of God manifest according to His will, Hebrews 4:2 proclaims the truth of the matter: "*For unto us was the gospel preached, as well as unto them: but the word preached did not profit them, not being mixed with faith in them that heard it*" (KJV). Whatever has been sown in the heart and isn't subsequently watered and nourished with the Word of God will fail to grow in faith, for "*faith comes by hearing and hearing by the Word of God*" and will consequently cause the promises of God to be of no effect. And if the Word of God hasn't taken root, something with an earthly magnetism will, which will reduce God's own magnetic ability to draw them near to Him.

In terms of spin waves' involvement with a substance's ability to respond to a magnetic field, while all materials respond to some extent to a magnetic field, only permanent magnets, like Ferro magnets, actually are capable of generating a magnetic field themselves. What differentiates Ferro magnets from other materials is that the atoms in substances like iron and nickel will spontaneously set themselves in order, like a well-trained army, one behind another, line by line, all pointing in the same direction. This unity of focus enables those with magnetism to draw other material with magnetic potential toward them. The readiness to respond in this way differentiates a Ferro magnet from other material. A paramagnet, on the other hand, is an example of a metallic substance with magnetic potential, but its atoms are usually in a random state, each pointing in different directions. In order to bring about this substance's magnetic potential, however, it usually simply requires a "training session" with a permanent magnet.

Having their atoms in order and all pointing in the same direction—which is the physical condition that enables a permanent magnet's magnetic field to generate an electric field—permanent magnets need only gently sweep over the paramagnet's surface. The resulting electric field behaves as the trainer in this context, providing the paramagnets with the ability to emulate or copy the path of order. Following the Ferro magnet's training—an example of a second order phase transition—the paramagnet's atoms line up, which results in it too being capable of magnetically drawing others with magnetic potential near. However, there is one vital difference. While a Ferro magnet's magnetic ability is limitless, a paramagnet needs to be connected to the permanent magnet in order to draw others to itself.

A good example of this is a demonstrable occurrence with a Ferro magnet and paper clips. If the Ferro magnet connects to a paper clip, then that clip can connect with another, which can connect to another, and so on down the line. As long as the first one is connected to the Ferro magnet, the others will adhere to the magnetic connection. However, if that first one departs from the Ferro magnet, then all of the paperclips will just fall to the ground and remain alone. This provides an illustration of why we must draw near to God Himself, because counting on others' connection to God for our magnetic connection to Him leaves us vulnerable to falling and, ultimately, alone.

Being prepared and lining up, at least in one's heart, is of vital importance. It is in this way that we may understand order. Concerning order, Job 33:5 says this: "*If you can answer me, set your words in order before me; take your stand.*" In other words, speak words that line up with your intentions as well as your commitments, and don't succumb to any persuasion to get yourself to veer off course. Two chapters later it says, "*Teach us what we should say to Him; for we can prepare nothing because of the darkness*" (Job 37:19). The polar opposite of light was apparently stealing the ability for Job to speak effective words, words that lined up with his desire instead of his reality. And in Isaiah it says, "*Thus says the Lord: 'Set your house in order'*" (Isaiah 38:1). All of these Scriptures mention order, and while only two explicitly speak about order in the context of words that are spoken, a fuller knowledge of spin waves should make applicable Isaiah 38:1 as well.

As mentioned, spin waves are disturbances in magnetic material that affect the ordering of the atoms and the electrons within them. Magnetism, in turn, occurs when the atoms in a substance line up; so in essence spin waves disrupt the ordering potential. What is important, therefore, is first to discern the character of any magnetic field, and second, to maintain magnetic connection with a genuine magnet, or better yet, the true global magnet Himself. A true permanent magnet, like a true shepherd of the sheep, is in a state of constant preparation, training the sheep to hear the Master's voice and His words. Opposed to the spin waves of strife, this permanent magnet remains in peace and sweeps over those who will hear, who will ultimately set their house in order.

Second Timothy 2:24-26 presents the sentiment this way: *"And a servant of the Lord must not quarrel but be gentle to all, able to teach, patient, in humility correcting those who are in opposition, if God perhaps will grant them repentance, so that they may know the truth, and that they may come to their senses and escape the snare of the devil, having been taken captive by him to do his will."*

God and His Word mirror in every way an ideal permanent magnet, and completing the illustration, the training session most perfectly relates to the Law as spoken of in the Old Testament. Galatians 3:23-24 says, *"But before faith came, we were kept under guard by the law, kept for the faith which would afterward be revealed. Therefore the law was our tutor to bring us to Christ, that we might be justified by faith."*

The Law was given to Moses as a type of permanent magnet that would sweep over the Israelites and train them in the way that they should go. *"And these words, which I command thee this day, shall be in thine heart: and thou shalt teach them diligently unto thy children, and shalt talk of them when thou sittest in thine house, and when thou walkest by the way, and when thou liest down, and when thou risest up. And thou shalt bind them for a sign upon thine hand, and they shall be as frontlets between thine eyes. And thou shalt write them upon the posts of thy house, and on thy gates"* (Deuteronomy 6:6-9 KJV).

Israel was being prepared by God to fill their hearts and minds with the Word of God, so that they would know His voice and develop their faith. And in reminding Israel of God's provision, Moses provided an exhortation to obedience: *"Man doth not live by bread only, but by every word that proceedeth out of the mouth of the Lord doth man live"* (Deuteronomy 8:3 KJV). The Law was provided for instruction as well as for training, just as a magnet sweeps over a material with unrealized magnetic potential.

As all men have been born into a family that has received life, which is the light of men (John 1:4), they have also been provided a measure of faith. As Romans 12:3 verifies: *"God has dealt to each one a measure of faith."* Conclusively, all men have been enabled with such magnetic potential. However, due to the many and various variables that are encountered in life, the opportunities to shun the power of the true permanent magnet (God) are as plentiful as the stars in the cosmic galaxies. The words of the wise in this world—those who have been able to discern many truths by virtue of their own hard work—unfortunately often discourage the notion of God, intellectually judging such a belief as simple-minded folly. And their words of unbelief constitute a local magnetic field of force.

As mentioned, there are also many local magnetic fields that draw people toward them. When one truly considers the battle between these local magnetic forces and God's global magnetic force, the original situation in the garden of Eden may be seen in a new light. The Serpent in the garden of Eden thus represents a local magnetic force that drew Adam and Eve toward him by the use of flattery and deception. In essence, he led them to believe that they were more worthy than they were; in other words, he tried to generate the anti-fruit of pride. He led them to believe that God was holding out on them, that they truly deserved to have what God Himself had. Or in other words, Satan tried to generate jealousy and contempt. By using his words to generate these anti-fruitful thoughts he was, in essence, planting seeds that would grow when watered by similarly spirited words. The planted seeds would thus become receptors for any subsequent frequency that adhered to their nature, and this is at the root of what is required when magnetic connections are formed and have an opportunity to operate.

As magnetic connections operate according to the principle that opposites attract, everyone is consequently susceptible to becoming bound. What becomes apparent when applying this attraction to human beings is that in the same way that a negative charge and a positive charge attract, one person's apparent strength would appear to attract another person's apparent weakness. With either the desire to feel more capable by bonding with someone stronger or to feel one's strength or superiority by bonding with someone weaker, many bonds are not equally yoked, meaning in this context that the root intention behind the bond may be less than pure. Due to these potential attractions, therefore, one requires discernment.

First John 4:1-3 broaches this call to watchfulness: "*Beloved, do not believe every spirit, but test the spirits, whether they are of God; because many false prophets have gone out into the world. By this you know the Spirit of God: Every spirit that confesses that Jesus Christ has come in the flesh is of God, and every spirit that does not confess that Jesus Christ has come in the flesh is not of God. And this is the spirit of the Antichrist, which you have heard was coming, and is now already in the world.*"

All of this pertains to our thoughts and our ability to discern the character of God and to line our thoughts up with God's thoughts. This is the only way to resist the other magnetic forces in the world, i.e., "*the cares of this world and the deceitfulness of riches*" spoken about in Matthew 13:22. Our cares and concerns are like spin waves that decrease the magnetic power between us and God. And it all starts with our thoughts; whatever we think about is already taking shape as words within our own heads, and it's only a matter of time for those thoughts to manifest as words. But even if they are never actually spoken so that they produce sound waves, they are still voiced in the heart, and thoughts, like electrons, will generate an electric field.

In the event that the electric field does change, i.e., one's thoughts concerning a person, place, or thing change, there is a corresponding generation of a magnetic field. Attraction or repulsion characterizes this field, and it can be experientially verified that when someone's thoughts change about an issue, a resulting judgment ensues, which corresponds with this field's character; a corresponding like or dislike takes place. And if you purposely change what you are attracted or repulsed by, i.e., through empathy and understanding, there will be a consequent generation of an electric field, or a new wave upon which electrons or thoughts will travel.

Chapter 19
Separate Division

That people are scarcely aware of the infinite variables that determine a body's movement through time and space is proven most assuredly from the speed at which our technological advancements have taken shape. Every day there seems to be new acquisition of knowledge and, just as rapidly, a new application of that knowledge appears on the scene. While I am truly humbled by all of the knowledge that individuals have attained, no man at any time suddenly invented the principles upon which these new advancements operate. Rather, a new fundamental understanding is made possible due to the acquisition of a new knowledge base. It has, above all, required openness to new processes of thought. Unlike the cyclical nature of closed systems, where energy is conserved, open systems operate according to the steady influx and outflow of energy waves that flow between the system and its environment.

Learning and accepting new ways of thinking and perceiving requires this openness. While an open system can grow, a closed system remains unchanged. The concepts of openness versus closedness provide a basic illustration of a fundamental polarity that exists in the world. According to the theory of energy principles, proposed by psychoanalyst Carl Jung, along with entropy and equivalence the principle of opposites made a debut. According to one conceptual understanding, the principle of opposites serves the role of allowing us a way by which we may truly gain an understanding of a polar extreme. For instance, we never really know wealth until we have experienced poverty, or we really will never understand love until we have experienced hatred. While this may be an accurate assessment of this concept's experiential worth, I propose that its *true* value pertains to its initial placement in the universe.

Where did polarity first find its conceptual home? In the very beginning, right after light had been conceived and had broken through the darkness. As a testament to God's reaction to the light, Genesis 1:4-5 says, "*And God saw the light, that it was good; and God divided the light from the darkness. God called the light Day, and the darkness He called Night.*" Now of particular interest is the fact that at this point, God was not referring to the daily cycles that we refer to as *day* and *night*. This division was made the first day as Genesis 1:5b proclaims: "*So the evening and the morning were the first day.*" The separation between light and darkness that is controlled by the daily ascending and descending of the sun beyond our particular horizon is spoken of in Genesis 1:

> *Then God said, "Let there be lights in the firmament of the heavens to divide the day from the night; and let them be for signs and seasons, and for days and years; and let them be for lights in the firmament of the heavens to give light on the earth"; and it was so. Then God made two great lights: the greater light to rule the day, and the lesser light to rule the night. He made the stars also. God set them in the firmament of the heavens to give light on the earth, and to rule over the day and over the night, and to divide the light from the darkness. And God saw that it was good. So the evening and the morning were the fourth day (Genesis 1:14-19).*

This was done on the *fourth day*! The first separation mentioned, however, was accomplished on the *first day*.

The division between light and darkness that occurred on the first day is obviously referring to a completely different division. In validation of this assumption, if these were speaking about the same separation, why would the work be done twice? Did God make a mistake the first time? This would be particularly surprising in light of His tendency to stop and consider His work and "see that it was good." This would appear to be an unreasonable and faulty assumption. For God doesn't do the same thing twice, because He knows the beginning and the end of everything as Isaiah 46:10 says, *"Declaring the end from the beginning, and from ancient times the things that are not yet done, saying, My counsel shall stand, and I will do all my pleasure"* (KJV), and Revelation 1:8 confirms this: *"I am the Alpha and the Omega, the Beginning and the End."*

If He is the beginning and the end, then He would be in charge of birth processes as well as death cycles. As is evidenced by most things in the world, the second law of thermodynamics—entropy, whereby things naturally tend toward decay—applies to all living entities. However, as Prigogine and Stengers noted in their book, *Order Out of Chaos*, there is often a process that takes hold after the process of decay, where the decaying substance develops new structures and evolves in a new way. Referred to as dissipative structures, contrary to any assumption that this might be unsupported by the Word of God, this conceptual framework happens to dovetail perfectly with God's way of operating in the universe. Here is an example of a process wherein evidence of these structures would be present:

> *The hand of the Lord came upon me and brought me out in the Spirit of the Lord, and set me down in the midst of the valley; and it was full of bones. Then He caused me to pass by them all around, and behold, there*

were very many in the open valley; and indeed they were very dry. And He said to me, "Son of man, can these bones live?" So I answered, "O Lord God, You know." Again He said to me, "Prophesy to these bones, and say to them, 'O dry bones, hear the word of the Lord! Thus says the Lord God to these bones: "Surely I will cause breath to enter into you, and you shall live. I will put sinews on you and bring flesh upon you, cover you with skin and put breath in you; and you shall live. Then you shall know that I am the Lord."'" So I prophesied as I was commanded; and as I prophesied, there was a noise, and suddenly a rattling; and the bones came together, bone to bone (Ezekiel 37:1-7).

According to the above Scripture, not only do God's ways support the concept of dissipative structures, which in essence has been an evolutionary concept, but we also see that by mentioning prophesying as a means by which this is accomplished, words play a vital role in the passage from death unto life. And if God's ways align with evolutionary processes, then He wouldn't make things twice, but would rather establish a bridge between two or more elements that would tie a physically evident process with a spiritual pattern. And this would be consistent with God's ways of operating consistently as Malachi 3:6 says, "*For I am the Lord, I do not change,*" a well as being a creator as Revelation 21:5 says, "*Behold, I make all things new*." Accordingly, Genesis 1:14 and Genesis 1:5 speak of two separate divisions, with the preliminary division on the first day emulating the heavenly division between God's power and Satan's.

Chapter 20
Polar Forces

Ask, and it will be given to you; seek, and you will find; knock, and it will be opened to you. For everyone who asks receives, and he who seeks finds, and to him who knocks it will be opened" (Matthew 7:7-8). As electrons flow from the negative pole to the positive pole in an electrical circuit, and we may assume that God is negatively charged due to the role that He serves in forming the salt of the earth with people, we may understand the importance of our role in asking, seeking, and knocking; otherwise His electrons (thoughts) will likely not flow in our direction. This pattern specifically relates to how God wants us to be drawn to Him. While He loves everyone, He is fundamentally bound to function according to the preestablished electrical dynamics, where as a negatively charged entity His fundamental electrical power draws toward Himself those who are ready and willing to succumb to His wisdom and love. However, in light of the magnetic aspect of this field, this implied sequence in the bonding process would be difficult to empirically measure because, like two magnets whose north and south poles meet, the bond occurs instantaneously.

The character of God must not be minimized with regards to this connection; He is the ultimate gentleman, and accordingly He will not push Himself on anyone who does not want Him in their lives. He has, however, enabled a mediator who is ultimately responsible for the connection. As an electrolytic capacitor in a battery conducts alternating current between a negative terminal and a positive terminal, so too does Jesus, the Word of God, as 1 Timothy 2:5-6 proclaims: *"For there is one God and one Mediator between God and men, the Man Christ Jesus, who gave Himself a ransom for all, to be testified in due time."*

It would appear that Jesus is the absolute battery, acting as both a primary as well as a secondary device. Initially when He is accepted as Lord and Savior, He serves as a primary converter of the chemicals of the "*water by the word*" (Ephesians 5:26) into electrical power. Secondarily, He serves the function of both a collector as well as an emitter of electric charge. And the character of this electrical power must not be ignored, for love is the essence of the power that He, operating as a battery, provides. This electrical power thus serves as the force that enables a positive light source to reach into the heart and be emitted from the eyes, to which Proverbs 15:30 alludes: "*The light of the eyes rejoices the heart.*"

This Scripture illustrates the light of Jesus, the Word, as an energy force that is capable of first converting the *chemicals* stored in the Word (Jesus) into light energy, which is subsequently absorbed and then reemitted through the eyes. When the Word is truly heard by a listener, it becomes absorbed and can operate as a seed in fertile ground. The end result is faith, as Romans 10:17 proclaims: "*So then faith comes by hearing, and hearing by the Word of God.*" Faith may then be said to be equivalent to all of the nuclear power within a seed, which like the vaporous phase, is brimming with unseen potential.

One who truly listens to and absorbs God's Word and accepts God's love and presence in their lives will then be furnished with the ability to emit the same power. As "*God is love"* (1 John 4:8), love is the essence of this electrically charged force field. And as it says in 1 John 4:17, *"Love has been perfected among us in this: that we may have boldness in the day of judgment; because as He is, so are we in this world.*" The Word of God, in this position, allows the energy from God's power to be placed in us, but only through Jesus; He is the mediator, acting in a similar capacity to the electrolytic fluid within a battery, that through Him we also are enabled to become like Him, a Mediator who transmits the power of the Word.

With consideration of the frequencies that we ourselves emit, which we must recognize as a fundamental aspect of the entire process with which we are intimately involved, that which we listen to corresponds absolutely with the words that we will speak. In turn, the words we speak are the very elements that propagate through space, with the sound-emitting frequencies travelling through matter and corresponding light frequencies being absorbed, reflected, or refracted by objects in the environment. Coupled together, these frequencies affect us as well as others.

With reference to the wave nature of electromagnetic frequencies, one aspect of waves becomes abundantly clear: When waves come into contact with other waves, there is interference. Consequently, when the waves that are associated with our words go forth and interact with other people's light emissions that also assume a wave form, both constructive and destructive interference may occur. The end result of these interferences is either an increase in amplitude of the combined wave, causing excitement, as in constructive interference, or a decrease of the wave's amplitude or a torrent of chaos, bringing about the effect of being drained or confused, as in destructive interference.

Understanding that the positive amplitude (crest) of a wave may be considered as one polar extreme, and the negative amplitude (trough) as the other may help us to identify polarity as an interfering aspect of God's universe. In one sense the dynamic of interference can be seen as a force that seeks to destroy, and in another sense it can be seen as a force that seeks to join and become one with the waves with which it collides. This is, in essence, the human plight. People, who emit both light and sound waves, are not only vulnerable to the waves emitted from others in their environment but are also vulnerable to resonant frequencies that abide in the earth's atmosphere. We are, therefore, caught very much in the middle in the sense that without knowledge of how to combat interferences, our lives become subject to undesirable results.

While terms such as positive and negative are essentially meaningless, as their value can only be discerned contextually, the results of any interference can be either beneficial or destructive. In reality, everything can be esteemed as either good or bad, depending not on the thing itself but the situation as well as on the hearts or intentions that are coupled with the propagating waves. These, in turn, are mere extensions of the frequencies that we have admitted into our lives; this provides substance to the notion that we do have choices in this world. Joshua 24:15 mentions one such choice: "*Choose for yourselves this day whom you will serve, whether the gods which your fathers served that were on the other side of the River, or the gods of the Amorites, in whose land you dwell. But as for me and my house, we will serve the Lord.*"

These particular choices are indicative of the same kinds of choices that people today have to make. Will they bow down to the things that their parents have taught them? Will they look around at the world in which they live and what everyone else is doing and succumb to that pressure? Or will they consult their own conscience and heart and serve the Lord? The choices that are ultimately made will have come to fruition as a result of the waves and frequencies (light and sound) that we admit into our lives. And these will determine any subsequent resonation, for we will be "in phase" with these waves.

There is no doubt the varying ways in which interference can play out in our lives has a tremendous impact on us as well as on others. As an example, when we meet someone who we automatically like and with whom the conversation flows, we can readily understand this interaction by recognizing how constructive interference is playing a vital role. The crests and troughs of our emitted frequencies/waves are in phase, or in other words they are lined up and flowing with each other. As a consequence, just as in a water wave, the amplitude of the resulting wave's crest would increase, which is experientially verified by an increase in excitement and joy. Conversely, when we are interacting with someone who doesn't share any of our interests and the conversation doesn't flow but feels like one is pulling teeth instead, we can readily recognize destructive interference as that which is occurring; when two waves are out of phase when they meet, the deadening effect is tangible.

These two particular illustrations, however, seem to indicate a correspondence between constructive and positive/good and destructive and negative/bad. This, however, is not an accurate assumption. Illustrations that provide a tearing down of this assumption would then be when two people with similar temperaments are really angry with one another. Due to their similar temperaments as well as similar state of mind, they would be considered *in phase*. However, the constructive amplitude in this case would be anything but positive. As an example to disrupt destructive interference's potential association with something undesirable, if one were to introduce a person who was *out of phase* with one of the individuals in the last illustration, one of the potential outcomes could be a calming down of the propagating frequencies.

This last example corresponds with one of the ways in which Jesus, the Word, is able to affect our lives; He (the Word) has a calming effect:

> *The Lord is my shepherd; I shall not want. He makes me to lie down in green pastures; He leads me beside the still waters. He restores my soul; He leads me in the paths of righteousness for His name's sake. Yea, though I walk through the valley of the shadow of death, I will fear no evil; for You are with me; Your rod and Your staff, they comfort me. You prepare a table before me in the presence of my enemies; You anoint my head with oil; my cup runs over. Surely goodness and mercy shall follow me all the days of my life; and I will dwell in the house of the Lord forever (Psalms 23).*

When we implicitly trust God and we believe that what He has said aligns with ultimate truth, we are free to walk in love, for as 1 John 4:18 says, *"Perfect love casts out fear, because fear involves torment.*" And love and fear thus provide another example of polar extremes.

Galatians 5:22-23 speak about the fruits of the Spirit, where love tops the list: *"But the fruit of the Spirit is love, joy, peace, longsuffering, kindness, goodness, faithfulness, gentleness, self-control. Against such there is no law."* Now, these fruits of the Spirit indicate one polar extreme. However, because this world operates according to laws of conservation, where ideally polar extremes will cancel out one another's *charges*, these represent only half of the powers that are in operation, as we are all too well aware.

The domain space where these polar forces exist is alluded to in Ephesians 6:12: *"For we do not wrestle against flesh and blood, but against principalities, against powers, against the rulers of the darkness of this age, against spiritual hosts of wickedness in the heavenly places."* Please notice that people, a.k.a. *flesh and blood*, are *not* our enemy. We can convince ourselves that they are, and why is that? Because we can't imagine that anyone with a conscience could do what some of them do. While there are some who have no conscience, i.e., sociopaths, most who are governed by this polar force are simply deceived into believing that a lie is the truth; the anti-fruitful frequencies have organized or trained the neurons in their brains to assume a closed loop connection that is incapable of being open to the truth. Second Timothy 3:13 speaks about the generation of these frequencies: *"But evil men and imposters will grow worse and worse, deceiving and being deceived."* While they persecute you, they are the real victims, being deceived as they are.

Adhering to an assumption that a fundamental polarity governs the systems in God's universe—if the fruit of the Spirit as mentioned in Galatians 5:22-23 is love, joy, peace, longsuffering, gentleness, goodness, faith, meekness, and temperance, governed by a force field—then a polar opposite field must exist. This opposed field then would be comprised of anti-fruits that oppose these specific fruits. In this case, the anti-fruits would be fear/hatred, sorrow, impatience, harshness, wickedness, unbelief, pride, and indulgence. However, while the fundamentals that govern the relationships between these fruit forces do indeed adhere to particle/anti-particle dynamics in that they operate from polar opposite domains, they more precisely function according to wave dynamics and their corresponding interference factors.

According to normal particle/anti-particle dynamics, when anti-particles collide, they annihilate one another with a resulting production of radiation. This is the basis of why the baryon asymmetrical universe doesn't make sense. Matter and anti-matter, just like a particle and its anti-particle, are predicted to cancel each other's *charge* upon collision. The evidence of light and heat having been in existence in the very beginning would have been accounted for by anti-particle collisions; however with consideration of our substantive universe, this hypothesis just doesn't work. In the same way that matter and anti-matter didn't cancel each other out in the beginning, so too the fruits and anti-fruits of the Spirit do not cancel each other out when they collide or interact.

When an anti-fruit and a fruit collide, the interaction corresponds with wave interference factors and specifically destructive interference. Whenever a quantity of a fruit of the Spirit (wave) collides with that of an anti-fruit, they are *out of phase*; the troughs and crests align as opposed to crest and crest or trough and trough aligning, as in constructive interference. The resulting wave will have diminished the individual identifying waveforms after collision, with the fruits of love and peace, in particular, tending to affect an interference that will produce a calmer wave.

This concept is exemplified in Mark 4:38-39: *"But He [Jesus] was in the stern, asleep on a pillow. And they awoke Him and said to Him, 'Teacher, do You not care that we are perishing?' Then He arose and rebuked the wind, and said to the sea, 'Peace, be still!' And the wind ceased and there was a great calm."*

Jesus' inward fruit of peace was emitted as a wave that traveled as a sound wave, coupled with the light of the Word of God, and it destructively interfered with the tumultuous waves of the sea. He also, however, had faith to believe that His words had power.

Once armed with knowledge about the physics of the Word, it becomes our responsibility to choose how we will be affected by the atmospheric frequencies, as well as how our own light emission will affect the world around us. First Peter addresses this: *"Therefore humble yourselves under the mighty hand of God, that He may exalt you in due time, casting all your care upon Him, for He cares for you. Be sober, be vigilant; because your adversary the devil walks about like a roaring lion, seeking whom he may devour. Resist him, steadfast in the faith, knowing that the same sufferings are experienced by your brotherhood in the world. But may the God of all grace, who called us to His eternal glory by Christ Jesus, after you have suffered a while, perfect, establish, strengthen, and settle you"* (1 Peter 5:6-10).

Experientially speaking, the factors that appear to affect a change in the polar domains, simply stated, may be connected to resistance, steadfastness, obedience, and an open heart to God's Word. These are the doings that seem to enable freedom from the effects and bonds that would so easily weigh us down, as if gravitationally pulling us toward the earth. These factors appear to be the only paths to true freedom, where we may rise above the storms of life. There is no one who escapes the trials that are faced by humanity. But we have been provided with the *"sword of the Spirit, which is the Word of God"* (Ephesians 6:17), and we can use that to resist the Devil, who primarily attacks our thoughts. *"For as [a man] thinks in his heart, so is he"* (Proverbs 23:7).

The way we think determines our perceptions, our interpretations, our hopes, ambitions, and dreams. The way we think in our heart may then, for better or for worse, describe our essence, or what spirit (resonant frequency) is in primary control of our life. However, with the foundation of the Word of God, which has been provided us, our perceptions and interpretations, as well as all other conscious processes can instead line up with the Word of God, or with truth instead of just facts that have been presented in a certain light. The option has been given us: *"For though we walk in the flesh, we do not war after the flesh: (For the weapons of our warfare are not carnal, but mighty through God to the pulling down of strong holds;) casting down imaginations, and every high thing that exalteth itself against the knowledge of God, and bringing into captivity every thought to the obedience of Christ; and having in a readiness to revenge all disobedience, when your obedience is fulfilled"* (2 Corinthians 10:3-6 KJV).

Our thoughts are thus equivalent to electrons in that in the same way that free electrons are vulnerable to the incomplete atoms that enter their zone, our thoughts, when they are not bound with the Word of God, are equally as vulnerable. Our thoughts, in essence, will determine not only that to which we are bound, but also all interferences that we will endure. When our thoughts become bound to the Word of God, whose alternate identity is Jesus Christ (John 1:14), however, not only can we bring peace and rest to an otherwise stormy situation, but we can also become the true *salt of the earth.*

Chapter 21
Particle-Wave Duality

While in the context of "*Draw near to God and He will draw near to you*" (James 4:8) God behaves as a permanent magnet, the electric aspect of His electromagnetism cannot be undervalued. As Jesus is said to be the light of men in John 1:4, Jesus and the Father God are accounted as one in John 10:30, "*I and My Father are one.*" God is light, and operates in the twinkling of an eye with both the speed and nature of lightning. As light, therefore, whatever applies to the phenomenon of light applies to the nature of God. With consideration of the wave-particle duality of light, therefore, we may understand Jesus' role as operating according to a photon, the mediating particle of the electromagnetic force. This is supported by the fact that Jesus has been called the Mediator of the new covenant in Hebrews 12:24, so we may validate this function.

The Father God's function, on the other hand, may be seen as one behaving like an electromagnetic wave. The particle-wave duality of light would thus be accounted for with the entire electromagnetic spectrum pertaining to the nature of God. In addition, that no one has ever seen God but has heard His voice suggests that He has reserved the frequencies that are undetectable to the outward senses to speak with those who love Him and who have within themselves the Spirit of truth.

Providing an account that appears to validate this claim, in 1 Kings 19:11-12 an angel of the Lord spoke to Elijah: "*And he said, Go forth, and stand upon the mount before the Lord. And, behold, the Lord passed by, and a great and strong wind rent the mountains, and brake in pieces the rocks before the Lord; but the Lord was not in the wind: and after the wind an earthquake; but the Lord was not in the earthquake: and after the earthquake a fire; but the Lord was not in the fire: and after the fire a still small voice*" (KJV).

The powerful manifestations of thunder and lightning are likewise expressions of both God's voice as well as His presence, as Revelation 4:5 declares, "*And out of the throne proceeded lightnings and thunderings and voices: and there were seven lamps of fire burning before the throne, which are the seven Spirits of God*" (KJV). However, these expressions simply rest on one end of the electromagnetic continuum, with the other end reaching all the way into the heart of mankind. The stillness of God's voice speaking to Elijah corresponds with the fruit of the Spirit, gentleness, as spoken of in Galatians 5:22: *"But the fruit of the Spirit is love, joy, peace, longsuffering, kindness, goodness, faithfulness, gentleness, self-control. Against such there is no law."*

The suggestion that God's true identity is electric light has its root in the fact that any manifestation of light, with its corresponding heat—which, like lightning affects air's compression, even within the body—manifests with accompanying vibrations in the form of sound. Herein lies the foundational origin of God's voice—light. God the Father, therefore, may be understood as the wave form of true light. Jesus, in turn, may be understood as the particle aspect of light that acts as a Mediator.

In the same way a photon has the ability to infiltrate the electron shells in an atom and affect a resonant event, Jesus (the Word of God) is able to infiltrate the heart and mind of man. If a person has the Word of God and/or Spirit of God within the heart, there will be a consequent resonation with the light. If not, the light will travel on until it finds a heart in which it can find rest. God's ability to draw people who have their inner heart lined up toward His magnetic pull, as well as His wave-particle duality that enables His light to be absorbed and consequently reemitted, portrays God as the force after which the force of electromagnetism is known to operate. According to this assumption, the proposal that lightning may have accounted for the initial expansion of the universe, spoken about in the big bang theory, provides additional substance to God being the initiator of the universe. With lightning regarded as operating within the domain of the electromagnetic force, would lightning not be considered one of God's primary means of expression?

From knowledge gained scientifically, lightning is considered electromagnetic radiation that occurs in response to the buildup of electric charge in the atmosphere. Essentially a gigantic spark that heats the air to around 54,000 degrees Fahrenheit, lightning's high temperature of heat naturally causes the air to expand. This, in turn, creates a shock wave that is heard as a booming sound wave, or thunder. Interestingly enough, the ice crystals and hailstones within the clouds are the substances that enable the lightning spark to develop in the first place. While the ice crystals have a positive charge, the hailstones have a negative charge; in the environment of a storm, an updraft pushes the ice crystals to the top of the cloud and a down draft pushes hailstones to the bottom. Because of the consequent separation of charges, the negative charge at the bottom of the cloud becomes attracted to the positive charge of the earth's ground, and this attraction registers in volts of electrical power (voltage or potential difference) and lightning is produced.

A question comes to mind: Could it be that the "*treasures of hail*" spoken about in Job 38:22 find their full expression of worth in their involvement in the production of lightning? Hail would thus appear part and parcel of enabling God to send forth His thunder in response to His light.

While the electromagnetic force is comprised of both electric and magnetic fields that are perpendicular to one another, operating separately but in conjunction with one another, both fields are generated by electric charges. With the manifestation of lightning attached to the sound of God's voice (see Revelation 4:5) revealed as thunder, which is simply a response to light's atmospheric effects, God's nature would appear to align Itself more closely to the electric field. Accordingly, if a Cartesian graph were to be drawn, where a vertical line runs perpendicular to a horizontal one, God's power would appear to abide on the vertical axis with the magnetic field perpendicular to it.

As the earth produces its own magnetic field, being polarized with both a north and south pole with its core as its generator, we may see the earth's horizontal plane, upon which waves travel, as the X-axis. Considering that the earth, as opposed to the heavens, is so characterized, the other magnetic forces that compete for our attention come into fuller view. They are quite literally all that is earthly—all that draws us away from the vertical axis, the electrical axis, the axis on which God's ultimate power abides.

Regardless of the structural relationship between the two, however, the electric and magnetic fields are intrinsically connected, with both being tied to electric currents that are formed by the motion of electrons in atoms. What becomes evident, therefore, is that electrons are at the root of an atom's state of charge, which in turn propels them toward or away from objects in their environment. Making our hair stand on end when our bodies sense danger or creating a magnetic response of attraction or repulsion, electrons make contact with the electrical signals in our environment. While photons (light) are also involved, it's the excess or scarcity of electrons within an atom or molecule that is responsible for the charge as well as for the potential connections to which a body may become bound. By extension, we may again understand thoughts playing a similar role in our lives, as they do appear responsible for our connections.

As the mediating particles of the electromagnetic force, photons instigate the movement of electrons through the energy levels within an electron cloud. When a photon's energy and an electron's energy resonate, the electron is able to absorb the light energy which enables it to pass to another level within the atom. This is the physical pattern that parallels the Word of God as a *receptor seed* that consequently enables one to *hear* and gain faith *"by hearing, and hearing by the Word of God"* as Romans 10:17 declares. Photons, however, can do more than simply allow electrons to move more freely within an atom; they can actually liberate it from the atom entirely.

When an electromagnetic frequency is fast enough, such as ultraviolet light, a photon can loosen an electron from an electron shell. Because this is often accomplished by ionizing radiation, the action can more precisely mirror theft. However, if the light is characterized as belonging to the life which is in the Word of God as John 1:4 declares, it has the capability of supplanting unfruitful thoughts (electrons) and severing them from our cognition. Similar to how an electron requires all of a photon's energy to be absorbed in order to be liberated, a person often requires saturation in the Word of God in order to be freed from their bondages. To substantiate this, according to John 8:32, "*If ye* ***continue*** *in my word, then are ye my disciples indeed; and ye shall know the truth, and the truth shall make you free*" (KJV).

Chapter 22
Lasers

When we are imprisoned or in bondage, light's electron-loosening power pronounces liberty and freedom as Proverbs 10:22 says, *"The blessing of the Lord makes one rich, and He adds no sorrow with it."* However, a greater blessing than having wrong thoughts captured and taken away would be to join in a bond with the Word of God, have God's thoughts become yours, and become the salt of the earth. Ionizing radiation, the photoelectric effect, oxygenation, and the light that is emitted from a laser all have the power to loosen electrons (thoughts) from their energy shells, but none of them have the power to produce a neutralized salt. Therefore, while we can see these effects as potentially harmful, as thoughts being stolen might readily parallel mind-debilitating diseases such as dementia and Alzheimer's, the loosening of thoughts that bind in self-deprecating loops can certainly be understood as a polar opposite process with promise.

By and large, this loosening ability appears to align with the keys of the kingdom, which serves a universal and critical role in establishing the ways of God's kingdom. Therefore, with respect to these keys more than any other electromagnetic manifestation, the light power of a laser appears to emulate that which binds as well as that which loosens. The keys of the kingdom of heaven that were given to Peter in Matthew 16:19 were the keys of binding and loosing that *"whatever you bind on earth will be bound in heaven, and whatever you loose on earth will be loosed in heaven."* While these keys were given to Peter because he had received revelation knowledge from the Father that Jesus' true identity was the Son of God, the ability to receive this knowledge had required faith in Jesus' words and in their intrinsic power. Hebrews 4:12 mentions their might: *"For the word of God is living and powerful, and sharper than any two-edged sword, piercing even to the division of soul and spirit, and of joints and marrow, and is a discerner of the thoughts and intents of the heart."* Lasers, therefore, probably more than any other manifestation of light aside from lightning, provide a true look at the Word of God, particularly with regard to their ability to bind as well as to loose.

Paralleling the reflective qualities of God's white light of love, lasers utilize two reflective glasses within their optical resonating cavity. As light passes through the laser, it reflects back and forth between the two mirrored surfaces and becomes amplified in the process until it is emitted. With respect to a laser's capability of performing the function of the keys of the kingdom, a laser's light has been proven to be able to bind together as well as to loosen delocalized electrons in metal, in the processes of welding and cutting. With the illustration of a laser pertaining to God's Word, it also becomes apparent that the *gain medium* that lasers require might indeed represent the body of Christ. As unity of the body of Christ is something that God desires in order for it to be considered *one* body under the headship of Christ, a gain medium consists of a large group of atoms that provide the necessary energy in order for atoms to be transformed into their excited states.

While the gain medium in a laser can be in any of the transitional states—vapor, liquid, solid, or plasma—what is called for among other things is purity. What a laser beam requires is for the light to be coherent, a state in which the photons are logically ordered and the electromagnetic waves are in phase. Reminiscent of what occurs in a Ferro magnet that is able to draw a substance with magnetic susceptibility toward itself, in order to occupy a select position under the headship of Christ and fulfill the completion of the body under Him, the white light of love would appear to be a thread in this process, as Ephesians 4:15-16 declares: *"But, speaking the truth in love,* [we] *may grow up in all things into Him who is the head—Christ—from whom the whole body, joined and knit together by what every joint supplies, according to the effective working by which every part does its share, causes growth of the body for the edifying of itself in love."* Being aligned with and governed by the source of power, therefore, whether magnetic or electric, appears to be a necessary condition for the operatives of a laser, a magnet, or the body of Christ.

As in a laser that emits light with power, it would appear that in order for God's children to also emit light with power, their minds and thoughts must be transformed and *excited* by the light of the Word of God: *"And do not be conformed to this world, but be transformed by the renewing of your mind, that you may prove what is that good and acceptable and perfect will of God. For I say, through the grace given to me, to everyone who is among you, not to think of himself more highly than he ought to think, but to think soberly, as God has dealt to each one a measure of faith"* (Romans 12:2-3). If every member's thoughts were lined in order toward God's Word, where they would not only be drawn toward Him magnetically, but would also emit light due to having absorbed the Word with remittance of light as a result, they would individually "*stand perfect and complete in all the will of God*" (Colossians 4:12), and corporately they would emit laser light-like power.

Chapter 23
Resistance

As no pattern in heaven or manifestation of the pattern on earth occurs in isolation but may be understood as intricately tied to others, one specific connection between electric and magnetic fields corresponds with André-Marie Ampère's circuital law. Discovered in 1826, the law specifically relates to the generation of a magnetic field when an electric current travels within the confines of a closed loop. We understand that when there is a change in an electric field, as with thoughts, this change generates a magnetic field, as with feelings, and when there is a change in the magnetic field, this change likewise generates an electric field. As Ampere's law focuses on the magnetic fields that are generated by electric currents, using the pattern of this law it would be conceivable to determine how thoughts generate human passion and emotional feelings. And under the condition that a direct current, as opposed to an alternating current, was traveling in the closed loop, one might be able to determine the emotions that generated the thoughts. In light of this potential, it appears feasible to consider people in these *closed loop* terms.

In addition, when an electric current travels in a wire according to its natural flow, if a magnetic compass approaches the wire, the compass needle will align itself perpendicular to the wire. If, on the other hand, the current is reversed through manipulation, the compass needle will again be perpendicular to the wire but will be facing in the opposite direction. This suggests that whenever there is manipulation and/or redirection of a natural electrical flow (stream of thought), there is a corresponding magnetic repulsion as a response. This is experientially validated when we consider people's tendency to resist new ideas that don't flow according to the closed loop track of their mind, when they feel they are being manipulated, or when the message doesn't correspond to their expectations or to their desires.

In the context of a closed loop, the current, or thought wave, is not free, but it is instead bound and susceptible to magnetization or polarization. This appears to be the condition of people who are easily led astray by local magnetic forces in the environment. Thoughts that are bound and not free appear to travel in closed loop circles, and instead of being able to reasonably defend a positional attitude, which would suggest *openness*, their travels are bound within a *closed* circle. As a consequence, local magnetic forces may easily gain power.

With the properties of attraction and repulsion, these forces can then affect a polarization or division between those who are so controlled. When we consider bound currents that flow in one direction in a closed loop, there is a consequent separation or division into two oppositely charged poles with an area in between. Normally, according to electromagnetism, opposite charges attract, but due to the distance between the charged poles, they do not connect but rather resist attraction. This *in-between* area, therefore, may be seen as charged with a resistant energy, which consequently represents a great divide, where it could be said that life's struggles exist.

Analogous, in fact, to this gap between two poles is the "firmament" spoken of in Genesis 1:5-7: *"And God saw the light, that it was good; and God divided the light from the darkness. God called the light Day, and the darkness He called Night. Then God said, 'Let there be a firmament in the midst of the waters, and let it divide the waters from the waters.' Thus God made the firmament, and divided the waters which were under the firmament from the waters which were above the firmament; and it was so.*" This division alludes to a greater division than the one addressed Genesis 1:16-17, which says, *"Then God made two great lights: the greater light to rule the day, and the lesser light to rule the night. He made the stars also. God set them in the firmament of the heavens to give light on the earth."*

The division between day, marked by the sunlight, and night, marked by the moonlight, and the apparent separation between waters—earthly waters and heavenly waters—and a firmament in between would appear to indicate that the firmament existed between two poles. Following this assumption, there would be an energy in this firmament marked by what may be termed a magnetic moment that points from the south pole to the north pole of a magnet, which could also be understood as simultaneously resisting the opposite direction. Magnetic moments are generated by the motion of electrons that have formed electric currents. As the rotation of electrons (thoughts) comes from both spin and orbital motion, these rotations cause a separation between charges; while spinning, the negative charge of the electrons responds to its own charge with repulsion and in response undergoes polarization, with two magnetic poles of equal magnitude but opposite polarity being formed.

A question may be posed: Perhaps the implication in Matthew 6:28 that human beings "spin" is simply in reference to the fact that we, like other polarized bodies, are vulnerable to being divided within ourselves, with our thoughts spinning around within our minds and repelling each other. Romans 7:15-19 addresses this human dilemma: *"For what I am doing, I do not understand. For what I will to do, that I do not practice; but what I hate, that I do. If, then, I do what I will not to do, I agree with the law that it is good. But now, it is no longer*

I who do it, but sin that dwells in me. For I know that in me (that is, in my flesh) nothing good dwells; for to will is present with me, but how to perform what is good I do not find. For the good that I will to do, I do not do; but the evil I will not to do, that I practice." But God, in His merciful understanding of our condition, takes pity on us, and instead of looking at the flesh has decided to look upon the hearts and has accounted us worthy—if only we believe in His Son Jesus Christ and honor *Him*.

It is within this polarization of mind and body, with the mind of Christ (see 1 Corinthians 2:16) as the optimal Mediator, that the resistant energy manifests itself most fully. Galatians 5:22-23 speaks of the fruits of the Spirit: "*But the fruit of the Spirit is love, joy, peace, longsuffering, kindness, goodness, faithfulness, gentleness, self-control. Against such there is no law.*" As we have said, however, due to polarization, the fruit of the Spirit does not exist in isolation, free from the field of resistant energy.

An anti-fruitful field led by the anti-fruits of fear and unbelief or rather fearful and unbelieving thoughts (electrons) seeks to attach themselves to any and all unsuspecting Adams. First Peter speaks about the root source of these anti-fruits: "*Be sober, be vigilant; because your adversary the devil walks about like a roaring lion, seeking whom he may devour. Resist him, steadfast in the faith, knowing that the same sufferings are experienced by your brotherhood in the world*" (1 Peter 5:8-9).

While neither the fruitful nor anti-fruit domain is entirely free from being resisted—due to the second law of thermodynamics, entropy, and its consequent nature of tending toward decay—anti-fruits, which are closer to our selfish instincts, require a lot more work to resist. As a logical consequence of existing on the path of least resistance, anti-fruits have the greater advantage. Just think of a time when you've been angry or even sorrowful; the feelings attached to anti-fruits of the flesh naturally desire expression and often require much restraint to silence their voice or quench their force.

The comparison of thoughts with electrons finds further support with the consideration of streams of electrons (electric currents) in and out of a closed loop. When thoughts run rampant, they do tend to gather and follow a stream, and they do gain strength like an electric current. The more we think along one track, the more our emotions become involved, and the consequence of having our emotions involved is that we can no longer be impartial. These one-track thoughts then affect our state of mind so that the momentum of the thought current causes a response of repulsion or attraction, which then becomes evident by the words we speak and light energy that we emit.

That our emotions can often get us in trouble is not a question; all rash comments that fly out of our mouths are caused by flaring emotion. And feelings of self-pity and sadness over our losses, although often valid, do nothing but steal our joy and peace and keep us bound in a pit, so to speak. Emotions, therefore, may be considered as the domain where a magnetic moment occurs in a human being—that moment when thoughts spawn a magnetic response, aligned in an order facing either left or right toward a field of either attraction or repulsion.

A magnetic moment in these terms is then considered to be the moment when an electrical response in the form of thoughts transitions into a magnetic response registered by the emotions. In consideration of this structural pattern, the electromagnetic force might be readily understood as two phases of a wave, much like a crest and a trough represent the phases in a water wave. With regards to the intrinsic power that appears to be stored in a resistant energy domain, the difference between the apex of the crest and the lowest point on the trough may be readily understood as equal in force to potential difference or voltage.

Chapter 24
Potential Difference

While potential difference signifies the difference between the electrical potential of two points in an electrical circuit and thereby defines a circuit's ability to do work, this concept is not restricted to occurrences in electric circuits. Think about the difference in power that waves of a raging storm have compared to a relatively calm sea. In the first example, the difference between the apex of a wave's crest and the lowest point of its trough is large and carries a commensurate amount of energy to a strong electric volt. In a relatively calm sea, however, a similar difference would be very small, equivalent to little power.

To illustrate potential difference in a mechanical system, one may use a simple example of lifting a boulder to the top of a hill. When the rock is on the ground, it is considered to be at its ground state; however, when it is moved up the hill an equivalent amount of energy to the amount exerted transfers to the boulder, and the boulder has now absorbed that energy in the form of potential energy. To figure out the potential difference or mechanical power, i.e., kinetic energy, one need only subtract the initial energy from the energy stored as a result of having had work performed on the system.

One implication of potential difference as a concept is that it may conceivably be applied to any living system which undergoes a transition, under the assumption that change requires an amount of effort. While the concept of voltage harnesses electrical power, other examples may simply register as a *force*. One thing is certain—living systems confront resistance. Within these systems, a point, which would represent the same kind of energy resistance that occurs in a pendulum that fluidly transitions between kinetic and potential energy, becomes apparent; the momentum of one form of energy alternately overrides the resistance of the other. While almost imperceptible, the point of transition from one energy form to another would appear to harness the greatest potential of the system, equivalent in power to a magnetic moment, or in real terms, the moment of choice.

There appears to be a point, similar to the point in all transitional phases where "the rubber meets the road," so to speak—a moment of physical friction or resistance, wherein the choice to act according to principles or fleshly instincts comes into play—and we may understand the source of any resistance as embodied in the fruit of the Spirit or the anti-fruit of the flesh. This is the point, coincidentally, where words as well as actions begin to manifest. An equivalent concept has been presented by Prigogine and Stengers in *Order Out of Chaos*, where there is a point in self-organizing systems at which a choice between going up or down, left or right appears to exist; while the choice of which way the system goes has been accounted for by variables such as gravitation, I might propose that magnetic fields are involved. Nonetheless, a correlation can be drawn between this and God's way of operating in that God's ways of operating go against, or resist, the natural flow of this world, with this world having fallen into the dominion of the antichrist spirit: "*For all that is in the world—the lust of the flesh, the lust of the eyes, and the pride of life—is not of the Father but is of the world*" (1 John 2:16).

Due to an association between this world and the magnetic field that emanates from its core, what becomes evident is a necessity to resist certain magnetic forces, which may well correspond to the anti-fruit of the flesh. Closely associated to this idea is the property of diamagnetism, which is the tendency to oppose an applied magnetic field, which similarly appears in all materials that are affected by the electromagnetic force. Whereas magnets induce a positive magnetic moment, diamagnetic materials induce a negative moment. When an external magnetic field is applied to a diamagnetic material, the current loops that the electrons have created, which in turn create a magnetic field, align in a way that opposes the applied field. Diamagnetism, in this regard, is in essence equivalent in value and force to electrical impedance.

As a measure of opposition to an electrical current's passage, according to LotK, electrical impedance may be understood as a disrupter of the electrical current through which the Spirit of God might flow. It's important to realize though that in light of the many and varied electrical frequencies available, electrical impedance may also serve as a disruptor of other electrical frequencies, and can therefore act in the same capacity as a valve that opens and closes and allows or forbids entrance of certain frequencies. If the door to God's Word has been opened and remains a dominant force, then the anti-fruits are more easily resisted and not allowed passage; if, on the other hand, doors to resonant frequencies that connect to antichrist spirits have been opened, the force of resistance will be in operation against the fruit of the Spirit, impeding access to these instead. Revelation 3:20 addresses this concept: "*Behold, I stand at the door and knock. If anyone hears My voice and opens the door, I will come in to him and dine with him, and he with Me.*"

As opening a door implicitly involves resisting *not* opening it, having faith requires resisting unbelief. Similarly, embracing a state of unbelief also involves resisting faith, and thereby hope. "*The good fight of faith*" (1 Timothy 6:12) then requires a coupling of faith and anti-unbelief, just as falling away from God or purposely never coming to Jesus requires a coupling of unbelief and anti-faith. These couplings add strength to a particular path's direction. Interestingly, while the pattern of this mechanism adheres to the theoretical framework of quantum chromodynamics, which adheres to the strong force, there are aspects of the couplings that belong more accurately to the weak force.

In the same way that an electron and a positron act as stabilizing forces for the identities of a neutron and a proton, respectively, during the process of beta decay (a weak force process), anti-unbelief and anti-faith in the above examples act as stabilizers for faith and unbelief, respectively. In order for faith to maintain its identity, anti-unbelief must be present as a stabilizing force; in order for unbelief to maintain its identity, anti-faith must act as a stabilizer. When the stabilizers are present, those that require stabilization are able to function according to their potential, whether good or bad. Extending from this conceptual framework, what may become evident is an almost imperceptible resistant force that operates at the stabilizing level, such as anti-faith resisting hearing the Word of God, to ensure that faith doesn't come, or anti-unbelief resisting unbelief and opening the door to hearing the Word of God, so that faith may have a chance to come. While both processes involve resistance, only resisting unbelief requires work on account of the natural flow of decaying processes, and because of this work, the energy utilized is stored as potential energy.

With regard to resistance, if the following concept were to be illustrated using waveforms, one might say that the point of greatest potential resistance abides at the point in the center of a wave's crest or trough. These two points in a wave designate where the weight of choice appears to be the greatest; while the point on the crest must resist falling or must choose where to fall and resist the other direction, the point on the trough resists fighting to get out of its well. Again, there's an almost imperceptible resistant force that keeps us from breaking free and/or keeps us going in circles, and overcoming this resistance may be seen as the fight that is set before us; it's subtle and deceptive and it entails being sober and vigilant, "*because your adversary the devil walks about like a roaring lion, seeking whom he may devour. Resist him, steadfast in the faith, knowing that the same sufferings are experienced by your brotherhood in the world*" (1 Peter 5:8-9).

Chapter 25
Conductivity

The principle of electrical conductivity is patterned after the graceful mode by which Christ both gave gifts to men as well the mode by which those same gifts allow others to receive of God's power. According to Ephesians 4:7-8, "*But to each one of us grace was given according to the measure of Christ's gift. Therefore He says: 'When He ascended on high, He led captivity captive, and gave gifts to men.'*" And in verses 11-15, it says:

> *And He Himself gave some to be apostles, some prophets, some evangelists, and some pastors and teachers, for the equipping of the saints for the work of ministry, for the edifying of the body of Christ, till we all come to the unity of the faith and of the knowledge of the Son of God, to a perfect man, to the measure of the stature of the fullness of Christ; that we should no longer be children, tossed to and fro and carried about with every wind of doctrine, by the trickery of men, in the cunning craftiness of deceitful plotting, but, speaking the truth in love, may grow up in all things into Him who is the head—Christ (Ephesians 4:11-15).*

With the consideration that electrical conductivity is concerned with the ease by which electrons pass through a medium, we may understand the transmission of grace that is coupled with Christ's gifts to men in a similar way. The ability to receive anything from God through Jesus Christ requires faith, which comes through the avenue of grace which is a gift from God. This grace, in turn, is provided to the body for one primary reason: It is given to men so that they may conduct His love, His Spirit, and His Word to the world. While electrons (thoughts) require an ordering of themselves the way atoms in general need to be ordered if they are to become magnetic, interestingly enough the most important thing appears to be abiding in a state of delocalization, or in other words in a state of liberty or freedom. When electrons are in this state they are not bound to certain energy levels but are rather free to move throughout their own atom as well as move over into adjacent atoms.

Having the ability to lose electrons and form positive ions (cations) which are surrounded by delocalized electrons is what makes metals good conductors of both electricity and heat. We may see a parallel emerge if we consider that this loss of electrons emulates the process that God Himself undergoes when He joins with His people and enables them to become the salt of the earth. This is also corroborated in 1 John 4:17, which says, "*As He is, so are we in this world.*" While this Scripture is specifically alluding to God being love, as love is white light, by adopting His thoughts (electrons), the white light of His love, which is embedded in His thoughts, will be "*shed abroad in our hearts*" (Romans 5:5 KJV). The importance of the pattern of conductivity, therefore, is that by this process wherein thoughts (electrons) are free, the Word and the Spirit can go forth into "prepared" hearts without resistance.

That the electrons are free in a good conductor stands in sharp contrast to the electrons in atoms that don't conduct electricity or heat; these become bound to other atoms and become mixed as opposed to remaining pure. Being mixed in this sense relates to being "*unequally yoked together with unbelievers*" (2 Corinthians 6:14), as remaining pure has everything to do with righteousness. The second part of this Scripture asks the question, "*For what fellowship has righteousness with lawlessness? And what communion has light with darkness?*" Just as light and darkness cannot abide together in the same space, one whose measurement device is a right angle and must measure everything by it to make sure that it aligns to 90 degrees has a very different perspective than those who measure according to the theory of relativity.

While seemingly restrictive, such a measurement tool actually enables the greatest freedom; with everything aligning to a right angle, anxiety and fear that things are wrong and not right have little opportunity to take over. It's not unlike an invisible fence; once the parameters are known, the dogs within its confines have ultimate freedom to run wild. Such is the reality when God's Holy Spirit, complete with His fruit, runs freely throughout the body of Christ.

According to the heavenly pattern, therefore, electrical conductance appears to emulate the conductivity of both the Spirit as well as the Word. With the Holy Spirit represented as "living water" in John 4:10 and Jesus, the Word of God, mentioned in close connection with the conductive fluids of blood and water in 1 John 5:5-6: "*Who is he who overcomes the world, but he who believes that Jesus is the Son of God? This is He who came by water and blood—Jesus Christ; not only by water, but by water and blood. And it is the Spirit who bears witness, because the Spirit is truth,*" we may assume that the pattern of electrical conductance emulates these heavenly processes.

Accordingly, we may more fully recognize the process by which the ministry gifts were given to men as paralleling the pattern of electrical conductance. Jesus used the process to provide men with ministry gifts, which was used again to conduct the power of God to those who had been prepared to receive of the power. And in that these gifts are established by the Holy Spirit and the Word of God, they are gifts that have the potential to keep on giving. This potential, however, may remain in the potential realm or can multiply in a continual process of conductivity that emulates a pattern of seedtime and harvest; the Word may be sown in good soil; it may bear the fruit of righteousness, which in this context is the process by which power is conducted to others through the specific gifts that have been provided; or conversely, this process may be impeded by resistance factors.

According to the understanding that both the fruit of the Spirit and its anti-fruits abide at polar extremes and whose energy may be understood as abiding in an almost imperceptible resistant state, what becomes evident is that in order for conductance to occur, there must be the presence of a state of agreement between conducting elements. While a conductive metal requires delocalized electrons, which move freely within and between atoms, there is an implicit, and yet subtle, state of agreement between them; if there were not, there would be resistance. Nonetheless, regardless of this agreement, each delocalized electron (thought) is ultimately free and thereby alone.

The principle of superconductivity, however, occurs when electrons (thoughts) are coupled in pairs referred to as Cooper pairs, which are bound together at low temperatures. In this superconductive state, therefore, that is marked by almost zero resistance to electrical conductivity, there appears to be true agreement between electrons (thoughts). Having had the repulsion between each one's negative charge overcome by the attraction to the metal's positive ions, we may understand these positively charged ions, or cations, in terms of those who have been prepared to become the "*salt of the earth.*"

Perhaps of relevant importance is the reference to low temperature here, as it may very well pertain to one of the temperature conditions spoken in Revelation 3:15-16, "*I know thy works, that thou art neither cold nor hot: I would thou wert cold or hot. So then because thou art lukewarm, and neither cold nor hot, I will spue thee out of my mouth*" (KJV). With respect to this analysis, the superconductive condition that experiences no resistance to electrical flow would indeed appear to define a beneficial *temperature* condition for the body of Christ to inhabit. Thereby the gifts of the Spirit as well as its fruit would be able to be effectively disseminated throughout the world.

Above all, however, is that the white light of love is an inseparable part of this process of conducting the Word and the Holy Spirit. While the purpose of giving ministry gifts is so that the body of Christ may be edified and that it may fulfill its potential as Ephesians 4:13 states, "*till we all come to the unity of the faith and of the knowledge of the Son of God, to a perfect man, to the measure of the stature of the fullness of Christ,*" without love, this process is stunted. What is required for growth is honesty and as verse 15 of this same chapter says, "*but, speaking the truth in love,* [we] *may grow up in all things into Him who is the head—Christ.*"

The reflective quality of love, among other things, promotes the body to be honest with itself, and the Spirit bears witness of the Word spoken in Romans 8:1-2 that says: "*There is therefore now no condemnation to those who are in Christ Jesus, who do not walk according to the flesh, but according to the Spirit. For the law of the Spirit of life in Christ Jesus has made me free from the law of sin and death.*" And herein lies perhaps the greatest example of electrical conductance—the Holy Spirit, who inhabits the children of God on earth, bears witness with the Word, which is a record bearer in heaven, and imparts peace to the body of Christ. This process of bearing witness also may be likened to a magnetic moment, which determines the force that a magnet can have on electric currents, which makes ultimate sense when we consider God and His Holy Spirit as the magnet, and the electric currents as thought waves. Performing the function of a magnetic moment, therefore, the process of "bearing witness" is understood as the moment when the direction of God's magnetic pull corresponds with the direction of one's thoughts; what is in one's heart and mind and what courses through one's veins resonates with what the Father, the Word, and the Holy Spirit, as record bearers in heaven, have established and proclaimed.

Chapter 26
A Strong and Better Way

As higher pressure and lower temperature is related to the density of matter, the more bound we are to the flesh, the greater the battle will be that rages within us. Unless we are in a state where we are capable of transmitting signals—ideally without electrical resistance, as with superconductive power—this bound state signifies less freedom and increased pressure. While none of us are truly free from pressure, this does not correspond to God's perfect will for us. He has provided a better way, one where our hearts are changed so that we are not weighted down by the fight to resist but have entered into His rest. This rest, in turn, is a place where one trusts the love of God so much that there is no fear, as 1 John 4:18 states: "*There is no fear in love; but perfect love casts out fear, because fear involves torment.*"

Regarding the promise that such a "*place*" is available, Hebrews 4:1-3 says, "*Therefore, since a promise remains of entering His rest, let us fear [be observant] lest any of you seem to have come short of it. For indeed the gospel was preached to us as well as to them [Jews and Gentiles]; but the word which they heard did not profit them, not being mixed with faith in those who heard it. For we who have believed do enter that rest*" and Hebrews 4:9 repeats, "*There remains therefore a rest for the people of God.*"

God has therefore prepared a state of rest for His people to enter that doesn't refer to the rest available to those who enter the kingdom of heaven. With regards to what God promised to do when the fulfillment of Jesus' life, death, and resurrection came to pass, Ezekiel 36:26-28 provides us with this: "*I will give you a new heart and put a new spirit within you; I will take the heart of stone out of your flesh and give you a heart of flesh. I will put My Spirit within you and cause you to walk in My statutes, and you will keep My judgments and do them. Then you shall dwell in the land that I gave to your fathers; you shall be My people, and I will be your God.*"

This was spoken in the Old Testament before Jesus died for us, but Hebrews 12:1-2 alludes to that Spirit having already been provided and our consequent ability to resist without torment: "*Therefore we also, since we are surrounded by so great a cloud of witnesses, let us lay aside every weight, and the sin which so easily ensnares us, and let us run with endurance the race that is set before us, looking unto Jesus, the author and finisher of our faith, who for the joy that was set before Him endured the cross, despising the shame, and has sat down at the right hand of the throne of God."* The implicit message here is that if we *keep* our eyes on Jesus, on the Word of God, we may overcome the obstacles that present themselves along our journey's path.

While we are in this fleshly bound state, no matter our condition, none of us are exempt from the pressures and weights of this world, as Matthew 5:45 says, "*For He makes His sun rise on the evil and on the good, and sends rain on the just and on the unjust,"* and Ecclesiastes 9:11 says, "*The race is not to the swift, nor the battle to the strong, nor bread to the wise, nor riches to men of understanding, nor favor to men of skill; but time and chance happen to them all.*"

However, we have been provided with weapons: "*For though we walk in the flesh, we do not war after the flesh: (For the weapons of our warfare are not carnal, but mighty through God to the pulling down of strong holds;) casting down imaginations, and every high thing that exalteth itself against the knowledge of God, and bringing into captivity every thought to the obedience of Christ; and having in a readiness to revenge all disobedience, when your obedience is fulfilled*" (2 Corinthians 10:3-6 KJV). As this Scripture indicates, the battle resides in the mind, where our imaginations run wild with all the tormenting memories of what has happened before, as well as what could happen in the future.

As alluded to before, our obedience is fulfilled when we resist anything and everything that comes against the full knowledge of God and His Word. This consequently comes most often in the form of thoughts that are carried by our imaginations. When we are obedient, however, we will overcome our rampant imaginations as well as our knowledge of what is considered *wise* to the world and bring every thought in line with the Word of God. *"For the wisdom of this world is foolishness with God. For it is written, 'He takes the wise in their own craftiness'. And again, The Lord knows the thoughts of the wise, that they are vain"* (1 Corinthians 3:19-20). The implication of this Scripture is that when our thoughts don't line up with the Word of God, we are not only susceptible to being deceived, but our imaginations will affect a closed loop situation where we will be impeded from overcoming by faith.

According to 1 John 5:4-5, "*For whatsoever is born of God overcometh the world: and this is the victory that overcometh the world, even our faith. Who is he that overcometh the world, but he that believeth that Jesus is the Son of God?"* (KJV). Faith, which comes by hearing and hearing by the Word of God, overcomes! And most importantly, the Word of God enables us to overcome our imaginations. Our imaginations consequently represent a weakness, whereas God's Word represents a strength. In Isaiah 43:18-19 the Lord addresses our tendency to be caught up by imaginations, which are intricately tied to our memories: *"Remember ye not the former things, neither consider the things of old. Behold, I will do a new thing; now it shall spring forth; shall ye not know it? I will even make a way in the wilderness, and rivers in the desert*" (KJV).

As is spoken in John 14:6, Jesus is the "way": *Jesus saith unto him, I am the way, the truth, and the life: no man cometh unto the Father, but by me. If ye had known me, ye should have known my Father also: and from henceforth ye know him, and have seen him"* (KJV). Jesus, the Word, is the Truth, and is the *"Mediator between God and men"* (1 Timothy 2:5). The *"rivers in the desert"* in this case refer to the Holy Spirit as spoken about in John 7:37-39: *"If any man thirst, let him come unto me, and drink. He that believeth on me, as the scripture hath said, out of his belly shall flow rivers of living water. (But this spake he of the Spirit, which they that believe on him should receive: for the Holy Ghost was not yet given; because that Jesus was not yet glorified)"* (KJV).

Therefore, if we know Jesus, we know the Father, and we have been provided a way to overcome, a way by which we may "*fight the good fight of faith*" (1 Timothy 6:12). This way involves the usage of the weapons that have been provided us, that are intimately related to our thought life as well as to our heart's condition. But as it says in 2 Corinthians 10:4-5, the weapons of our warfare have a particular function: They are provided to us for the "*pulling down of strong holds*" and for "*casting down imaginations, and every high thing that exalteth itself against the knowledge of God*" (KJV). Primarily, these weapons help us train our minds, as a permanent magnet trains the atoms in a potential magnet (paramagnet) to align in an orderly fashion. And of primary importance is the usage of a *resistant force* that impedes the passage of certain signals while conducting the thoughts (electrons) attached to God's electrical signals.

Operating according to specific frequencies, the fruits of the Spirit and the anti-fruits are known, just as the elements are, by their spectral emissions. When the fruits of the Spirit are in dominion, the light emissions, no matter the color, have a reflective quality which corresponds to the reflectivity of the white light of love. Hereby we may understand how God knows the hearts of all men, as spoken in Acts 1:24. The more reflective our spectral emissions, the greater the resonance between God's Spirit and the receptor seeds within our heart. Consequently, the less reflective our spectral emissions are, the greater the chance that the receptor seeds within the heart resonate with the *dark* "light" frequencies of the air. This resonance, therefore, would correspond with the light that is emitted from a black hole that is heat-generated as opposed to light-generated.

As a black hole absorbs all electromagnetic radiation, as well as anything that approaches its *event horizon*, so too do these light emissions, which are in fact the frequencies upon which the anti-fruits travel. Instead of responding with reflection and empathy, this light responds with judgment and hatred instead, and as in an inelastic scattering process, it steals the light energy from others.

Closely tied to the treasure in our hearts, our spectral emissions are sensed by others as well as known by God. Matthew 6:21-23 alludes to this reflection or absorption: "*For where your treasure is, there your heart will be also. The lamp of the body is the eye. If therefore your eye is good, your whole body will be full of light. But if your eye is bad, your whole body will be full of darkness. If therefore the light that is in you is darkness, how great is that darkness!*" White versus black or light versus dark, therefore, has everything to do with the reflective or absorbing condition of the heart, which is revealed primarily through the eyes.

As members of the human family who are electrically charged and respond to magnetic attraction and repulsion, we are very well aware of our interactions with the fruit of the Spirit and its corresponding anti-fruits. We know when we have faith, when we fear, when we are joyful, and when we have sorrow. We know when we don't believe, when we are anxious, when we feel at peace, or when we have love for someone, and we all know the differences in our energy that result from all of these *heart* conditions. As God is love (white light), which is reflective of every spectral frequency, the other fruit will resonate with love's frequency just the way an electron resonates with a photon of a particular frequency and is often brought to a higher level. In an interaction with the anti-fruits, however, depending upon their level of absorption, the light, even pure white light, may well be entirely drained, as is experienced by a black hole. What becomes important then in using resistant energy to "fight the good fight of faith" is to resist the anti-fruit with the Word of God, which is the primary weapon of our warfare and belongs alongside the full armor of God spoken of in Ephesians 6:

"*Put on the whole armor of God, that you may be able to stand against the wiles of the devil. For we do not wrestle against flesh and blood, but against principalities, against powers, against the rulers of the darkness of this age, against spiritual hosts of wickedness in the heavenly places. Therefore take up the whole armor of God, that you may be able to withstand in the evil day, and having done all, to stand. Stand therefore, having girded your waist with truth, having put on the breastplate of righteousness, and having shod your feet with the preparation of the gospel of peace; above all, taking the shield of faith with which you will be able to quench all the fiery darts of the wicked one. And take the helmet of salvation, and the sword of the Spirit, which is the word of God*" (Ephesians 6:11-17).

Notice that the Word of God is our only offensive weapon, with the rest of the armor serving as protective gear. However, most things that come against us come up from the inside of us, which is why so many of us have recognized that we are our own worst enemy. But it's important to recognize that the principalities, powers, rulers of the darkness of this world, and spiritual wickedness in high places are the generators of the anti-fruit just as the Spirit of God is the generator of the fruit of the Spirit. And all of these operate within us, as Mathew 15:17-18 contends: "*Do not ye yet understand, that whatsoever entereth in at the mouth goeth into the belly, and is cast out into the draught? But those things which proceed out of the mouth come forth from the heart; and they defile the man. For out of the heart proceed evil thoughts, murders, adulteries, fornications, thefts, false witness, blasphemies: these are the things which defile a man: but to eat with unwashen hands defileth not a man*" (KJV).

The condition of our hearts, therefore, is contingent upon the thoughts that are generated, which in turn generate words as well as intent. In effect though, words become the outward display and expression of the dominant couplings in our lives, which can take the form of couplings between "fruit" and "(anti) anti-fruits" (as in faith and anti-unbelief) or anti-fruits and (anti) fruits (as in impatience and anti-patience. When we feel fear that is not in response to an outward threat, it serves no purpose other than to steal our faith or ability to love as God would. The generation of this kind of fear, therefore, is represented in John 10:10, as that with the sole intent of stealing, killing, and destroying: "*The thief does not come except to steal, and to kill, and to destroy. I have come that they may have life, and that they may have it more abundantly.*" As Jesus and the Father are one and "God is love" (1 John 4:8), we know that love counters this fear, which is supported in 1 John 4:18: "*perfect love casts out fear, because fear involves torment.*" In addition, in 1 Peter 4:7-8, we learn how we may escape this enemy's destruction: *"Submit yourselves therefore to God. Resist the devil, and he will flee from you. Draw near to God, and He will draw near to you."* Deductively then , we may assume that if thoughts generate words, then meditation on God's thoughts, which generate God's words, would then be the way whereby resistance of the devil is performed. And here we have an express depiction of the fruit of love versus its anti-fruit of fear.

Paralleling a pattern understood by quantum theorists, these fruit and anti-fruit couplings mirror the quantum chromodynamic theory of color charge. As color theory involves three primary colors—red, blue, and yellow (green)—so too does color charge. However, the term "color charge" only relates to the basic dynamics of color theory, where colors mix and create new colors, or in this specific case, mix to affect a neutrality of charge. A basic premise of color charge is that the charge of a quark affects the charge of another quark by neutralizing it, such as what occurs when complementary colors are mixed—mixing red and green, blue and orange, or purple and yellow results in a light-absorbing neutral brown, a color not present in white light, which is reflective rather than absorbent.

With the cancellation of charge as a basic goal, what occurs on the quantum level is that the force-carrying particle (boson) of the strong force, the gluon, carries its charge and deposits it on each of the three charge-carrying quarks within the nucleons (protons and neutrons). As a paint brush with a neutralizing color would affect a change to other colors, so too does the gluon; and its deposited color charge, which continually changes in response to the feedback that it receives from the differently charged quarks, creates the conceptual glue that holds the nucleons together on the deepest level.

In light of *LotK* theory, the fruit and anti-fruit of the Spirit may represent the interiors and exteriors of a gluon, which have been conceived to have color charge on the inside and anti-color charge on the outside. According to color charge theory, while the quarks in a proton (uud—up, up, down) or a neutron (ddu—down, down, up) can take on color values of red, green, and blue, the anti-quarks take on values of anti-red, anti-green, and anti-blue. And while the quarks are ascribed three color values along with three anti-color values, the gluons are mixtures of two colors, such as red and anti-blue, with color charge in the center like a nucleus, and the anti-color charge surrounding it, like an electron cloud. In the same way, *LotK* theory proposes that the fruit of the Spirit takes on values of love, joy, peace, longsuffering, kindness, goodness, faithfulness, gentleness and self-control that make up the core of most human beings, while the anti-fruits, which take on opposing values, surround them. This, of course, adheres to the perspective that people are born basically good, having been given the "light of men" by God and having been created in His image and after His likeness (see Genesis 1:26). The three color values of the quarks then may relate to the gifts—faith, hope, and love, as mentioned in 1 Corinthians 13:13: "*And now abide faith, hope, love, these three; but the greatest of these is love.*" These proposed gifts of the Spirit can then be understood as providing the "strong force" glue that abides in the heart of each person. However, as in the color-charge theory, there is also an anti-gift charge, and therefore the anti-gift charges of anti-faith, anti-hope, and anti-love have an equal opportunity of abiding in the heart. These gifts and anti-gifts then behave as stabilizing forces, for better or for worse.

And there need not be any more evidence of actual particles to substantiate this claim, for the light or darkness that powers these gifts, anti-gifts, fruits, and anti-fruits travels on a myriad of frequencies which are restricted by no boundary. Strongholds, consequently, occur in response to the percentage of correlated anti-fruit/anti-gift couplings.

If these interactions do not explain the myriad of frequencies as well as states available, particularly but not exclusively to people, another strong force interaction applies as well. While mesons (particles composed of quarks) are part of the hadron particle family, the pi mesons (the lightest mesons) successfully explain the low-energy properties of the strong nuclear force. Understanding the vitality of patterns, the pattern after which the pi meson antiparticles interact may shed light on the interaction between fruit and anti-fruit particularly.

In the same way that a pi plus meson ($\pi+$) is made up of an up quark and an anti-down quark, and a pi minus meson ($\pi-$) is made up of a down quark and an anti-up, creating a low-energy strong interaction, the fruit and anti-fruit may be patterned similarly—couplings of faith and anti-unbelief, joy and anti-sorrow, meekness and anti-pride, and love and anti-fear create a stronger platform of resistance than any fruit or anti-fruit alone. While the fruits and (anti)-anti-fruits (faith and anti-unbelief) create a strong contingency for God's ways, the anti-fruit and the (anti) fruits (sorrow and anti-joy) create one that adheres to the ways of darkness.

While under many circumstances, due primarily to a lack of solid commitment in mind and heart, these fruitful or fruitless couplings do not gain unbreakable strength and can be broken with conscious resistance, there remain certain couplings that resist being broken. These are the strongholds that are addressed in 2 Corinthians 10:4. All couplings, however, are difficult to break, which is why so many of us are bound; we are bound by the experiences of our past, and the fears that our past will be our future; we are bound by our judgments as well as by our interpretations of our own and others' worth. And the strongholds particularly relate to this last point. As the strong force is that which is experienced between the nucleons in a nucleus, which are the particles that determine the identity of an element, the force itself is intricately involved in making sure that someone or something does not become *loosened* from the holding identity; this is the area where strongholds take root.

Remember, when an element transforms into a new element, what occurs at the quantum level is that an up quark in a proton transforms into a down quark or a down quark in a neutron turns into an up quark. Through this process, the proton becomes a neutron or the neutron becomes a proton, and the end result is that a new element is created due to an element's identity being related to its proton count. This process is referred to as beta decay, either *plus* with the proton turning into a neutron, or *minus* with the neutron turning into a proton.

Beta decay is a process that is associated with the *weak force*, and due to its transformative capabilities is directly related to what occurs when Jesus is accepted into an individual's life: "*Therefore if any man be in Christ, he is a new creature: old things are passed away; behold, all things are become new. And all things are of God, who hath reconciled us to himself by Jesus Christ, and hath given to us the ministry of reconciliation; to wit, that God was in Christ, reconciling the world unto himself, not imputing their trespasses unto them; and hath committed unto us the word of reconciliation"* (2 Corinthians 5:17-19 KJV).

As with the process of beta decay, a mediating particle identified as the W boson, whose charge is associated with electron or positron emission or absorption, is necessary for nuclear transmutation, whereby an element assumes a new identity. Similarly, the Word of God is the energy carrier that enables a man to undergo a similar transmutation, and assume a new identity as a child of God.

What's of interest is that the weak interaction is an example of *symmetry breaking*, a process which describes how small fluctuations that act on a system at a critical point can determine the fate of the system. The critical point at which these fluctuations occur is equal to a bifurcation point at which systems appear to make a choice as to the path taken, as spoken about by Prigogine and Stengers. That the symmetry breaking point relates to a point where a system makes a choice—as well as appears to be the pattern after which a man can accept Jesus Christ into their lives and be renewed—comes to full light in Deuteronomy 30:19: *"I call heaven and earth to record this day against you, that I have set before you life and death, blessing and cursing: therefore choose life, that both thou and thy seed may live"* (KJV).

The critical point then pertains to the point at which a choice is made between left or right, up or down, blessing and cursing, and ultimately between heaven or hell. And these choices are seen as making an appearance first when we are given an opportunity to resist the anti-fruits corresponding to the fruit of the Spirit with the weapons of God's warfare and particularly with *"the sword of the Spirit, which is the word of God"* (Ephesians 6:17).

Chapter 27
The Graphite Principle

While words may be considered inert, due to the fact that they proceed out of the heart according to Matthew 12:34: "*Out of the abundance of the heart the mouth speaks,*" they consequently are charged with electricity. As we may be considered to be electrically charged bodies, waves that proceed from us will naturally emit a charge, which will affect and be affected by the surrounding electromagnetic field. Accordingly, words that are spoken will conduct either the Spirit of God or the spirit of the world, as John 3:31 declares: "*He who comes from above is above all; he who is of the earth is earthly and speaks of the earth,*" and Romans 8:5, "*For those who live according to the flesh set their minds on the things of the flesh, but those who live according to the Spirit, the things of the Spirit.*"

From these Scriptures what comes into focus is the comparison between fleshly and earthly, and Spirit and heavenly. With carbon compounds known to be the second most abundant element in the human body after oxygen, as well as the one that forms the basis of all life on earth, the difference between two allotropes of carbon come into focus. While diamond and graphite are composed entirely of carbon, the structural formation of carbon atoms within both substances is so different that diamond is the hardest natural substance, while graphite is one of the softest.

These two substances' divergent qualities shed light on how the inner structure of our hearts and thoughts can determine our essential function. This difference is most notable with regard to the potential reference to these formations in Matthew 21:42-44: "*Have you never read in the Scriptures: 'The stone which the builders rejected has become the chief cornerstone. This was the Lord's doing, and it is marvelous in our eyes'? "Therefore I say to you, the kingdom of God will be taken from you and given to a nation bearing the fruits of it. And whoever falls on this stone will be broken; but on whomever it falls, it will grind him to powder.*"

Verifying the association, if anyone, having the metaphorical structural composition of graphite, falls on diamond, he/she would be broken, but if diamond fell on graphite, the graphite would be ground to powder. This metaphor thus relates to Jesus being the diamond upon whom many will fall. When we willingly fall on Jesus though, we will simply be broken; like a broken vessel enables light to shine through the cracks, when we are broken the light of Jesus can more perfectly be revealed. If, however, we never willingly fall on Jesus, eventually on judgment day when "*we shall all stand before the judgment seat of Christ*" (Romans 14:10), He instead will fall on us and crush us to powder the way diamond would crush graphite. In application of this understanding, *LotK* proposes a new principle: The Graphite Principle.

As all metals can conduct electricity, graphite also has electrical conductive properties. However, due to its internal structure, where only three out of the four carbon atoms are delocalized or free, the conductance of electricity is limited to the planar field. Because of this internal structure, one out of the four carbon atoms remains bound and limits the graphite's ability to conduct electricity beyond a side-to-side movement. In metals with good electrical conductivity, however, conductivity extends beyond this level to include conductance in all directions. All-inclusive conductivity, therefore, requires that all electrons are delocalized, which represents complete liberty. The configuration of electrons in graphite, however, limits electrical conductivity and makes it incapable of sustaining electrical connection with electrical signals that come from above.

The relativity of the graphite principle to the condition of people is that this relates to the all-too-familiar occurrence of placing people above God. Instead of seeking God Himself and developing an intimate relationship with the Creator, people's relationships with God are contingent on their relationships with other people. Rather than having a relationship style that is patterned after a right angle, according to true *righteousness* by faith, where the directionality of the vectors is at both one-hundred and eighty degrees (a straight horizontal line) as well as ninety degrees (a right angle), people's relationship styles are limited to the planar field.

When the graphite principle is in operation, communication remains on this earthly level for those who can be crushed by the Rock, who remain bound to the earthly elements, and for those who question Jesus' authority. In speaking with his son, Solomon, David made a plea in 1 Chronicles 28:9: "*Know thou the God of thy father, and serve him with a perfect heart and with a willing mind: for the Lord searcheth all hearts, and understandeth all the imaginations of the thoughts: if thou seek him, he will be found of thee; but if thou forsake him, he will cast thee off for ever*" (KJV). Our thoughts and thus our hearts determine, above all, our inner structure and whether we will overcome or whether we will be crushed, whether we may be able to enter into His rest or whether we will remain bound.

We must, therefore, be discerning of the things which we hear, determining whether they line up with the Word of God or not. But first, we must know the

Word of God in order to verify whether the words we are hearing resonate with God or whether they connect to an antichrist frequency and are contrary to the peace of God. We must know the Word in order to conduct the Word, and we must have the Spirit in order to connect with the Spirit; when we have both, we then become capable of experiencing the *triple point* of God, where all three phases of God can simultaneously coexist.

Chapter 28
Strength in Weakness

Two out of the four scientifically identified fundamental forces essentially deal with the level of binding energy between protons and neutrons. Referred to as the *strong interaction* and the *weak interaction*, the comparative strength in the designated terms signifies that positive binding energy plays a greater role in the strong force, while negative binding energy may be seen as playing a greater role in the weak force.

While positive binding energy relates to binding entities together with a force that can only be broken by a greater force, such as what occurs during a fission event, negative binding energy is understood as the *loosing* energy that overcomes that bond. So while the strong interaction refers to the binding energy that ultimately holds the nucleons together within the nucleus at the quark level, the weak interaction, as has been mentioned, refers to the decay processes that occur at the quark level that ultimately determine an element's identity. Again, this latter process refers to the process by which a man may become a new creature in Christ Jesus as 2 Corinthians 5:17 says, and at the same time incorporates the process of loosing, which according to Matthew 16:19 is a key of the kingdom of heaven.

The positive binding energy referred to in the strong interaction, however, provides the mechanistic pattern after which strongholds adhere as referenced in 2 Corinthians 10:4-5: "*the weapons of our warfare are not carnal, but mighty through God to the pulling down of strong holds; casting down imaginations, and every high thing that exalteth itself against the knowledge of God, and bringing into captivity every thought to the obedience of Christ*" (KJV). What becomes evident from this is that the positive binding energy appears to be able to be *loosed* by the negative binding energy implicitly involved in bringing our thoughts in line with the Word of God.

This, therefore, would also constitute the process of loosing, and because it pulls down strongholds, it is also a weapon. This negative binding energy, therefore, is a weapon of our warfare, which again is mentioned in Ephesians 6:17: "*And take the helmet of salvation, and the sword of the Spirit, which is the Word of God.*" The Word of God is our primary weapon and has been given to us to pull down strongholds or overcome the strong bonds that exist at the thought level, which ultimately hinder a person's ability to become a new creature in Christ Jesus. Therefore one may conclude that the Word of God (Jesus) *looses* those things which have been bound, with Psalms 146:7 substantiating this in relation to a primary function of our Lord Jesus Christ: "*The Lord* ***looseth*** *the prisoners*" (KJV).

An interesting correlation appears to exist between the binding energy of the strong interaction and the weak force. While there are two sides to the binding energy coin, one likened to the binding involved with the process of fusion, and the other likened to the loosing involved in the process of fission, the fact that the energy required to loose is stronger than the energy required to bind is noteworthy. This loosing energy actually appears to parallel what occurs during the weak interaction when stabilizing particles, such as the electrons or positrons, are expelled from the system; they are *loosed* from maintaining their stronghold connection to the atom that is undergoing the process of renewal. With consideration of this connection, we may be able to see the strength in this apparently weaker force.

Where does this strength come from? While there is an implicit coupling of neutrons and electrons and protons and positrons that appear to enable these nucleons' identities, the creation of such a bound state between protons and neutrons actually requires a releasing of energy. As water releases heat in order to become ice, the process of positive binding is an exothermic process, releasing heat in order to transition into a more solid or bound form. Alternately, negative binding energy, loosing energy, is characterized as an endothermic process and requires the absorption of heat energy. Interestingly enough, it is this process that transforms a solid into a vaporous state that not only stores more potential energy but also opposes entropy, or the process of decay. The strength of life where potential is at its most abundant is thus equated to a more vaporous or spiritual state of being, which corresponds to a *point* when the *heat* of the Spirit is absorbed into the heart.

We may conclude, therefore, that the loosing aspect of binding energy is not only of greater intrinsic value, as it corresponds with the intents and purposes of God, but also that due to the loosing aspects of the weak force, the weak force may in fact be *stronger*. In addition, Second Timothy creates a case whereby the Word of God assumes a phase transitional state that is imbued with freeing potential, which implies strength due to potential's storage of energy: "*But the word of God is not bound*" (2 Timothy 2:9 KJV). And John 8:31-32 adds, *"If you abide in My word, you are My disciples indeed. And you shall know the truth, and the truth shall make you free."* From these Scriptures, we may therefore assume that the greater the potential energy, which comes as by *absorbing* the Holy Spirit of truth, the greater the probability that we, *the prisoners*, will be set free.

Chapter 29
He

As the Word of God (Jesus), as well as physical light, has been sent to loosen the prisoners from their bonds with darkness (see Psalms 146:7; Isaiah 42:7) and to provide them with a strength to overcome, we see the ability to overcome physical barriers as a physical property of the element helium. At a significantly cold temperature, liquid helium appears incapable of remaining bound. As the second most abundant element in the universe after hydrogen, helium (He), like other elements, undergoes phase transitions. However, at the frigid temperature of approximately 2.17 Kelvin, the normal fluid phase of helium-4 is capable of transitioning to a phase where there is almost no resistance to liquid flow, which means that it passes into supernatural territory.

Passage into this cold territory enables helium-4 to actually escape all confinement. For example, when placed in a container, the liquid helium will *creep* up the walls and out over the edge, or it will tunnel under any barrier that presents itself as an obstacle. Provided with similar nomenclature as superconductivity—another phase transitional property with the otherworldly capability of having no resistance to electrical conductance—this lack of resistance to liquid flow is referred to as a superfluid phase. Interestingly, this superfluid phase also mirrors the state in which polaritons exist—the couplings of infrared photons and phonons, which were mentioned in relation to the initial state of the universe.

While the strong interaction's positive binding energy between helium's nucleons (two protons and two neutrons) is extremely high, which could imply a greater density due to an intrinsic loss of energy in the creation of its bond, helium is the second lightest and most buoyant element in the universe. While hydrogen surpasses helium's buoyancy, it also happens to be flammable, which indicates the element's vulnerability to becoming mixed, as opposed to remaining pure; it was this flammability that caused the disastrous consequences for the LZ 129 Hindenburg in 1937, which ultimately called for its replacement with helium. Similar to how people, who are vulnerable to readily accepting the thoughts and words of the world, may become easily reactive and consequently emit explosive light energy, so too does hydrogen appear to explode in flames with certain environmental influences. Helium, however, has a propensity for remaining pure.

Having the identity of a noble gas on account of its complete electron valence shell of eight electrons and consequent lack of reactivity, helium's properties parallel those who have entered into a state of rest, as is mentioned in Hebrews 4:3: "*For we who have believed do enter that rest,*" and this appears to be a phase transition that continues to be available according to Hebrews 4:9: "*There remains therefore a rest for the people of God.*" Similar to the creation of a salt, which appears in a neutralized state after the restless oppositely charged ions become bound, the process of entering into God's rest parallels what Noah appears to have done as the eighth person to enter the ark. As a noble gas like helium is spared from reactivity due to its completed valence shell, so too does it appear that Noah's entrance into the ark enabled it to enter a state of rest.

With regard to paralleling the creation of a salt, while the ions that make up a salt become *mixed* and are therefore not *pure*, when bound together they enter a state of neutralized rest, like helium in its pure state. Helium, therefore, may be assessed as having similar properties to those of God Himself, and interestingly enough the elemental symbol of helium is *He*. A striking similarity also exists in that both God and helium seem to advance their supernatural activity at very cold temperatures. According to Psalms 147:16-18, the cold appears to be a fundamental condition for many of God's works: "*He gives snow like wool; He scatters the frost like ashes; He casts out His hail like morsels; who can stand before His cold? He sends out His word and melts them; He causes His wind to blow, and the waters flow.*" Only His wind appears to be able to overcome His cold.

That snow, ice, superconductivity, and superfluidity occur at very cold temperatures appears significant. While ice causes a separation of charges (a loosing) within clouds that ultimately results in lightning and thunder (potential manifestations of God's voice), superconductivity—which is a result of a binding of electrons in a metal, or Cooper pairs, also at very cold temperatures—allows for the transmission of electrical signals (resonant frequencies). While superfluidity allows for supernatural escape or supernatural loosing, snow, as well as the Word of God, appears to be the respective manifestations of the triple points of both water and God. There's something about the cold, and there's something about God's Word, that when sent forth is capable of causing supernatural transitions, as in the "*treasures of the snow*" (Job 38:22 KJV).

With regard to helium's ability to transition into a superfluid state, a comparison must be noted between helium-4's ability to make the transition and helium-3's ability to enter the same phase. The temperature at which the isotope helium-4 (two protons and two neutrons) makes the transition is higher than the temperature required by helium-3 (two protons and one neutron). In light of this finding, it became apparent that each molecule in helium-4 actually behaves as a boson particle, or as an energy carrier.

What's interesting about this is that in most atoms, neutrons and protons are considered fermions, particles that make up matter and obey their own statistical rules (the Pauli-exclusion principle) that pose limitations on these particles' ability to simultaneously share quantum states with other particles. Bosons, however, are known as energy carriers that adhere to Bose-Einstein statistics, which *allow* for the simultaneous occupation of states.

What does this mean? It means that a group of bosons may occupy the same space; as a photon, which is the energy carrier for the electromagnetic force, is a boson, this simultaneous sharing of states accounts for the ability of photons to gather together into a cohesive stream of laser light. And in the case of helium, its ability to escape any confinement signals that the particles have behaved as carriers of supernatural energy, which would mean that they obeyed Bose-Einstein statistics rather than the Pauli exclusion principle. In effect, helium-4 atoms have the ability to bypass any corporate limitations, such as those imposed on helium-3 that differs only in that it has one less neutron. Perhaps this is just another example of the wonders of the neutron. Already established as the key to nuclear power, it may also hold a key to the supernatural transitioning power of helium-4.

That superfluidity occurs at very cold temperatures not only corresponds with superconductivity, but also relates to a potentially uncomfortable environment for human beings. The significance of this relates to the fact that, first of all, when we are comfortable we tend to rely upon ourselves; we tend to have confidence in our own ability to provide for our own needs. In other words, we become bound to self-reliance. This, however, does not reflect God's perfect will. Because of His tremendous love for His children, He wants to shower them with blessings. However, without trust in His power to bring all of the energy carriers (bosons) together and create a coherent stream of energy that may overcome natural boundaries, people remain operating according to the rules for fermions that can only operate according to rules that apply to solid matter. As a result, people *become* incapable of overcoming and are overcome instead.

With reference to Psalms 147:16-18, a correspondence appears to exist between the required temperature for superconductivity and superfluidity and the manifestations of the supernatural power of God. Regarding the question, "*Who can stand before His cold?*" the implicit message is that God is to be respected and feared, for His *cold* represents, in this regard, not only a climate that is beyond human endurance, but is also equivalent in value to His Word, before which men are destined to stand…or fall. This, of course, is substantiated in John 12:48: "*He that rejecteth me, and receiveth not my words, hath one that judgeth him: the word that I have spoken, the same shall judge him in the last day*" (KJV).

While this last Scripture adds to our understanding of the roles of the Word of God, that among other things it is the final judge, verse 18 of Psalms 147 provides us with an illustration of how the Word is also used as an enabler in the sense that it enables an outside substance to absorb itself. For example, the process of sending the Word and having the end result be melted snow and ice would indicate that the Word that was sent had an endothermic effect on these phases of water. In other words, the spoken Word of God was equivalent to a substance releasing heat energy, which then was absorbed by the snow and ice, which brought them to a state closer to "freedom."

The illustration of God sending forth His Word, therefore, is provided to us as a clue as to how the Word operates; it is sent forth as sound waves into the atmosphere to accomplish the vision and will of the sender. That God sends forth His Word and melts the physical manifestations of His cold also implies that due to our natural inability to stand before such inclement weather, His ultimate desire is to care for His children's needs, as 1 Peter 5:6-7 expresses: "*Therefore humble yourselves under the mighty hand of God, that He may exalt you in due time, casting all your care upon Him, for He cares for you.*"

Specifically relating to the superfluidity that helium exhibits when it is able to creep up the sides of a container or tunnel through any barrier, God is capable of freeing the humble from their uncomfortable bonds as 1 Corinthians 10:12-13 suggests: "*Therefore let him who thinks he stands take heed lest he fall. No temptation has overtaken you except such as is common to man; but God is faithful, who will not allow you to be tempted beyond what you are able, but with the temptation will also make the way of escape, that you may be able to bear it.*" An ultimate truth is implicitly conveyed in these words: Besides warning that one maintain an understanding of one's ultimate vulnerability due to the fact that none of us is beyond temptation or sorrow, with God's help, with the fruit of the Spirit—faith and patience, as well as humility—we have the tools with which we are capable of overcoming the *wells* of life. God will not leave us nor forsake us (see Psalms 94:14), and He has enabled us to enter a state where we will be as helium and be provided with a way to escape our trials.

Adhering to the mechanism of superfluidity, supernatural *escape*, although not commonplace, is simply another operational platform that God has designed for the purpose of fulfilling His Word. With the fruit in operation, a confidence in the truth of God's Word, and an understanding of God's faithfulness, supernatural escape—based on the laws of superfluidity—remains on reserve for an appointed time. "*For with God nothing shall be impossible*" (Luke 1:37 KJV).

While at super cold temperatures, helium can transition into a new phase, acquiring a new property of superfluidity; it also registers as something with a different nature when it is released from radioactive material. Similar to how the simplest form of hydrogen has the identity of a proton, helium's two protons and two neutrons provide the helium nucleus with the alternate identity of an alpha particle; it is through alpha decay that an alpha particle is expelled. As a highly ionizing form of radiation, alpha decay occurs in elements with an atomic weight of more than 106, and it is through this process that electrons are readily extracted from atoms.

Accounting for much of the helium within the atmosphere, the process of alpha decay occurs in heavy atomic nuclei like uranium or thorium in response to the repulsion between the positively charged alpha particle (two protons plus two neutrons) and the rest of the nucleus, referred to as coulomb repulsion. While this certainly makes sense, assuming that this is patterned after a heavenly pattern, as everything appears to be, there is more to it than meets the eye. That helium is characterized by buoyancy and freedom, and an alpha particle is characterized as a highly ionizing particle, what becomes apparent is an implied fight over freedom.

As an ionizing particle readily extracts electrons, the alpha particle may be understood as behaving like a mind controller or a thought thief, i.e., a sociopath. Helium, however, like the angel that led Peter out of prison in Acts 12:7-10, leads the way of escape to freedom and has no corresponding effect on the electrons or thoughts of other atoms/Adams; the thoughts of those who are freed remain true to their own identity. Verse 9 of this chapter speaks of Peter's mental state during this process: "*So he went out and followed [the angel], and did not know that what was done by the angel was real, but thought he was seeing a vision*" (Acts 12:9). While Peter's ability to differentiate the dimension of vision/dream from concrete reality was affected, the character of Peter's thoughts remained intact. Consequently, whether the variable of speed brings about more volatility or some other factor, the differences in effect despite the similarities in structure cannot be considered void of significance.

Assuming the identity of an alpha particle, the atomic configuration of two protons and two neutrons becomes a bound entity that—while capable of being stopped by a few centimeters of air or by skin—if ingested can lead to cellular damage. The fact that it can cause damage if swallowed indicates that its harmful potential is connected to the way in which it passes into a body's system. As the verb "to swallow" implies, that passage has occurred through the mouth, and words are the entities that pass through the mouth; there would appear to be a correlation. However, in this case, something else becomes evident; when words are *swallowed*, it implies that the substance of words, thought by the heart or heard by the ears, has become internalized, meaning that they have taken root in the heart and are perceived as *right.* And when seeds in the form of words or thoughts take root, these become *receptor seeds* that determine all future frequency receptions. Paralleling alpha particle damage, therefore, reception and absorption of thoughts and words that correlate with anti-fruitful frequencies have the consequent potential of changing fundamental structural patterns (such as DNA) of the system.

The question of what makes the alpha particle potentially harmful when expelled from a heavy nucleus is of interest. While the universe's helium supply is said to come from alpha decay, why does helium have a different fundamental nature than that of an alpha particle? It would appear that this difference may be discerned when we examine the psychological roots of freedom and bondage. While helium is regarded as buoyant and free, an alpha particle may be regarded as an expelled prisoner. If we examine the psychological states of people who may be likened to either helium or alpha particles, what would we find? When people have been born into freedom, like helium, they can use their natural abilities to rise above the weights that might so easily beset them (see Hebrews 12:1). When they are bound as in an atom of uranium, however, and they are suddenly expelled into freedom, having a mindset of having been bound, there is often a consequent fear of that freedom, and fear, as we know, counters love.

To support this, we may look at the high rate of recidivism among the prison population. As a consequence of fear, there is often anger, not only at those who imprisoned them, but also at themselves. And once prisoners, who have been incarcerated for a long while, are released, the sudden freedom and consequent choices and responsibility can be overwhelming; the fear of being out of control often sets in. This fear then turns to anger and hatred with the corresponding need to release the explosive energy. Add this with the fear of the freedom of choices set before them, and the safety of prison life, where choices and responsibility are minimal, becomes attractive. Herein is one psychological illustration of the process involved with the bound becoming free. The freeborn who remain free, however, do not tend to respond in this manner. But the bound or the free who have been bound do. An example of the psychological nature of the bound becoming free is evident in Exodus.

Several chapters in Exodus are devoted to providing an account of Moses' attempt to get Pharaoh to free God's people from bondage in Egypt. As God's people, the seed of Abraham, they had been esteemed by God as "freeborn." However, having been brought into bondage by Egypt, after years of slavery, their mindset had changed; they no longer saw themselves as freeborn but had a self-estimation of themselves as slaves. Exodus 14:11-12 provides an account of this consequent mindset: "*Then they said to Moses, 'Because there were no graves in Egypt, have you taken us away to die in the wilderness? Why have you so dealt with us, to bring us up out of Egypt? Is this not the word that we told you in Egypt, saying, "Let us alone that we may serve the Egyptians"? For it would have been better for us to serve the Egyptians than that we should die in the wilderness.'"*

People who have been enslaved or been bound do indeed find freedom to be a threat. Existentialist psychologists understand this, as freedom requires, above all, choice and responsibility. It is far easier instead to relinquish control as well as all sense of responsibility. Galatians addresses the ease with which people often return to a state of bondage, in this specific case of an understanding of God in terms of laws and tradition from which Jesus had freed them:

> *But when the fullness of the time had come, God sent forth His Son, born of a woman, born under the law, to redeem those who were under the law, that we might receive the adoption as sons. And because you are sons, God has sent forth the Spirit of His Son into your hearts, crying out, "Abba, Father!" Therefore you are no longer a slave but a son, and if a son, then an heir of God through Christ. But then, indeed, when you did not know God, you served those which by nature are not gods. But now after you have known God, or rather are known by God, how is it that you turn again to the weak and beggarly elements, to which you desire again to be in bondage? (Galatians 4:4-9)*

In this way, we may understand the difference in character between helium and an alpha particle. While an alpha particle represents the psychologically bound, helium represents the "*free indeed.*"

Throughout the Bible a distinction is made between the freeborn, represented by Isaac, who was a promised child for whom Abraham had to stand in faith with patience, and the bound child. This latter is represented by Ishmael, who was the son of Sarah's servant, Hagar, whose birth was a result of a moment of Abraham's doubting God's promise; in other words, he momentarily resisted faith in God's promise, an understanding of God's love, and entered the bondage of fear instead.

Similar to how a free helium particle differs from a bound alpha particle, a free person differs from one who is bound. As an alpha particle is expelled with ionizing energy that has the character of a thief, helium overcomes boundaries and escapes and often provides others with a lift. With the ability to tunnel through barriers and climb over walls in an act of overcoming or being victorious, helium epitomizes the pattern of the process that frees prisoners from their bondages to fear, just as faith in God's Word is able to do. *"For whatsoever is born of God overcometh the world: and this is the victory that overcometh the world, even our faith"* (1 John 5:4 KJV).

Overcoming barriers by faith, therefore, is the way for us to obtain freedom from being entrained to live in bondage; in other words, from having our intrinsic circadian rhythms align to external environmental factors, such as light. That light is included in the description of entrainment lends support to the suggestion that environmental *lightness* or *darkness* of a spiritual nature, implicitly related to the circumstances, objects, and/or conditions by which one is surrounded, is intricately tied to our freedom and our bondage.

With this understanding, we may see that an *environment* that emits light frequencies that correspond with the fruit of the Spirit will be characterized by more freedom, while those that correspond with the anti-fruit will more readily enslave us in bondage. Attaining a state of freedom thus requires resisting a state of bondage, just as attaining a state of bondage requires resisting freedom. Acceptance of the frequencies that correspond with either lightness or darkness, therefore, is intimately tied to the resistance of the opposite *pole's* force. At the root of it all, this resistance occurs at the level of beliefs and thoughts that either correspond with the frequency of the Word of God or with the frequencies of the world. "*Now the Lord is that Spirit: and where the Spirit of the Lord is, there is liberty*" (2 Corinthians 3:17 KJV).

As "God is love," we know that any resistance to the love of God comes from the anti-fruitful force of fear, and the bond between freedom and love is the *anti-bond* of the coupling between bondage and fear, as it says in Romans 8:15: "*For ye have not received the spirit of bondage again to fear; but ye have received the Spirit of adoption, whereby we cry, Abba, Father*" (KJV). The implicit message is that if we return to bondage, we will return to fear, with the consequent result that should we be released again, our nature will be more like an alpha particle than a helium nucleus that intrinsically operates according to the law of liberty.

Chapter 30
Liberty's Light

"There is no fear in love; but perfect love casts out fear, because fear involves torment. But he who fears has not been made perfect in love" (1 John 4:18). As bondage and fear go hand in hand, so too does liberty and love. As the fruit and anti-fruit of the Spirit correspond with certain light emissions, occurring as emitted waves released from those who are connected with them, we can easily see that as a fruit of the spirit, love emits a certain light. As a consequence of love's ability to empathize with the hearts of others, its light emission is equivalent to a reflecting white light. Accordingly, we may assume that love's opposite—fear—would appear as an absorbing darkness, like a black hole's. Able to absorb all objects as well as any electromagnetic radiation that reaches its event horizon, the center of a black hole actually gives a glimpse of the outer darkness mentioned in Matthew 22:13-14: "*Then the king said to the servants, 'Bind him hand and foot, take him away, and cast him into outer darkness; there will be weeping and gnashing of teeth.' For many are called, but few are chosen.*"

As reflecting light and absorbing darkness define the original universal polarity, we may see this outer darkness as pertaining to the darkness in the first division mentioned in Genesis 1:4: *"And God saw the light, that it was good; and God divided the light from the darkness. God called the light Day, and the darkness He called Night."* The King James Version accurately makes a vital distinction. In this verse, *Day* and *Night* are capitalized, whereas the reference to day and night in Genesis 1:14, which refers to the diurnal and nocturnal cycles, is presented in lowercase letters. Verifying the assumption that "*God is light and in Him is no darkness at all*" (1 John 1:5), verse 9 says, "*walk in the light as He is in the light*" (1 John 1:9). First Thessalonians 5:5 addresses the fact that people can reside on either polar extreme: *"You are all sons of light and sons of the day. We are not of the night nor of darkness. Therefore let us not sleep, as others do, but let us watch and be sober."*

First John 1:5-9 verifies the assumption that God is light, and *in* Him is no darkness at all, and as *"God is love"* according to 1 John 4:8, we know that the light that He emits corresponds with love in which no darkness may abide. This loving light that casts out fear then affords us the Spirit of adoption which counters the spirit of bondage to fear; like an adopted son, we can freely walk in a new identity, free of the fear of being found an imposter. *"If we say that we have fellowship with Him, and walk in darkness, we lie and do not practice the truth. But if we walk in the light as He is in the light, we have fellowship with one another, and the blood of Jesus Christ His Son cleanses us from all sin. If we say that we have no sin, we deceive ourselves, and the truth is not in us. If we confess our sins, He is faithful and just to forgive us our sins and to cleanse us from all unrighteousness*" (1 John 1:6-9). Herein lies a foundational outline of the dynamic between God, as white light, and man. White light illuminates the true nature of all objects on which it is cast.

With the analogy of God as white light, we can understand the dynamics that ensue when this light interacts with darkness. As white light, God illuminates the truth about us and ultimately exposes our very nature. Those who have been purified, or are willing to undergo the process, will suffer loss of self or pride, but they will not recoil from the light. Those who are not, however, having darkness within them will recoil and scatter or cause interference with the light's trajectory. In other words, they will avoid, steal, or absorb it. And if and when the light is absorbed, it will remain in the thief's heart in the same way electromagnetic radiation remains in the core of a black hole; while the light remains in the invisible realm, it can expel x-rays, radio waves, and other radiation into the surrounding space.

In essence then, this describes any emitted light that doesn't correspond with the love of God as essentially perverted, having its root in darkness. Matthew 6:23 provides a description of this light: *"But if your eye is bad, your whole body will be full of darkness. If therefore the light that is in you is darkness, how great is that darkness!"* These dynamics, therefore, expressly apply to how lightness and darkness respond to white light, as well as how light responds to darkness. As there is no fear in love, there is no darkness in white light, which exposes the true intents of the heart and is either accepted with humility or resisted with contempt. Why? Because *dark* vulnerability needs to rid itself of anything that causes pain, whereas *light* vulnerability understands that the light is God's love, as is alluded to in Hebrews 12:6: *"For whom the Lord loves He chastens, and scourges every son whom He receives."*

If we desire to accept the Spirit of adoption whereby we are enabled to become a son, we will regard the light's exposure as a gesture of love; we will understand it as a gifted guidepost that will enable us to overcome and be free. And when we overcome, by obeying God's gestures of love, we will be said to have achieved a level of freedom. And we will be commanded to *"Stand fast therefore in the liberty by which Christ has made us free, and do not be entangled again with a yoke of bondage"* (Galatians 5:1).

Psychologically speaking, a tremendous connection can be discerned between freedom, fear, love, and understanding. With consideration of the process of elastic scattering, when one particle hits another, the scattered particle bounces off in another direction. The fact that the particles never make a true connection but only makes surface contact is key to this connection. In relation to people's interactions with God (Light), using the eight ball illustration provided earlier, the cue ball would assume the identity of God, spinning white with absolute direction and speed, with the motive of jostling the balls (His children) out of their comfort zones so they might fulfill His will.

While being plunged into a side pocket or slammed against other balls so that they land where God wants *them* to land is not always painless, with an understanding of God's ultimate love for His children all of this becomes tolerable. The point is that while Jesus has love and compassion for us, He ultimately wants to get us to a place where we will fulfill *His* will for our lives—not our own will, but *His*. While the less love we have in our hearts, the more this message will cause us fear, the more love we have in our hearts, the greater is our freedom to simply love and trust Him.

As "*God is love*" and love operates as white light and Jesus was called the *light of men* (John 1:4), we see that just as the white light of love exposes the true nature of man, that light (Jesus), the Word of God, can also reflect every spectral emission that draws near. Hebrews 4:13-15 communicates it this way: "*And there is no creature hidden from His sight, but all things are naked and open to the eyes of Him to whom we must give account. Seeing then that we have a great High Priest who has passed through the heavens, Jesus the Son of God, let us hold fast our confession. For we do not have a High Priest who cannot sympathize with our weaknesses, but was in all points tempted as we are, yet without sin.*"

And as white light doesn't absorb any color, but reflects the true nature of the emitted light, those who do not war against the light will feel essentially understood. They will not feel vulnerable and naked before the light; instead they will feel that the light deeply comprehends their inner experience. This truly is the experiential manifestation of love, and expressly describes all that God is. "*But God, who is rich in mercy, because of His great love with which He loved us, even when we were dead in trespasses, made us alive together with Christ (by grace you have been saved), and raised us up together, and made us sit together in the heavenly places in Christ Jesus, that in the ages to come He might show the exceeding riches of His grace in His kindness toward us in Christ Jesus. For by grace you have been saved through faith, and that not of yourselves; it is the gift of God*" (Ephesians 2:4-8). God's grace is rooted in love, as well as a deep understanding of the fundamental spiritual poverty of human beings.

Many of us may be capable of comprehending a similar compassion. For example, when we are in the presence of a three-year-old child who doesn't understand all that we do, any lofty expectations we might have are replaced by tenderness, because we discern their *actual* capabilities and are able to love them just as they are. This is the true character of God—patiently accepting of His children in their weakness, with the hope that they will trust Him to provide them with His strength. We are, after all, His children, and like a good and faithful Father, God is willing to prop up us on His shoulders when the waters get deep, when we are fearful and weak. At the same time, however, He has extended an offer to surpass our title as children and become sons; this is an offer to those who will accept the call to maturity, where our positional weakness is understood, but where our hope in the three phases of God is strong.

"*For as many as are led by the Spirit of God, these are sons of God*" (Romans 8:14). "*But when the fulness of the time was come, God sent forth his Son, made of a woman, made under the law, to redeem them that were under the law, that we might receive the adoption of sons. And because ye are sons, God hath sent forth the Spirit of his Son into your hearts, crying, Abba, Father*" (Galatians 4:4-6 KJV). And as sons, we will no longer be children of disobedience, which signifies a bound existence, who walk after the flesh and not after the Spirit, but will rather walk in the light of liberty.

As God's light illuminates the truth within us, we may begin to see the connection between fear, love, and understanding. When we love, we provide a heart of understanding to others. When we fear, we become defensive. This state not only makes us not listen so that we will understand others, but we become offended by not being understood ourselves. And when we feel misunderstood, we often enter an area of striving to be understood. But 2 Timothy 2:24 says, "*The servant of the Lord must not strive; but be gentle unto all men, apt to teach, patient.*"

What occurs on a psychological level when we don't understand or when we feel misunderstood is that we feel out of control, which causes us to fear; when fear sets in, our feelings of self-worth, identity, and competence begin to wane. As was stated regarding how darkness responds to the light, as a consequence of this fear our human inclination is to eradicate the body of the anxiety-provoking feelings. When a person in our environment highlights our weaknesses or makes us feel inept, for instance, not only is the frequency and spectral emission of their inner light causing destructive interference, but they are causing us to question our own worth and competence. Feelings of inadequacy, in turn, have a tremendous ability to impact our state of equilibrium, sending us unconsciously into a realm where our defenses take over.

Wanting to protect us, our defense mechanisms fight to make sure that we see the world from a point of view that has no damaging consequences on our own self-perception. In their push to convince us of our own worthiness, our self-defense mechanisms take us down a darkened road where other people's faults come into full view, where we can no longer see ourselves for what we really are, and where others' faults become highlighted instead of our own. The end result is that we maintain our own feelings of self-worth by disparaging the worth of others. Fear, in this context, naturally results in repulsion of the source, which is manifested as hatred. Hereby we see the anti-fruit of fear as the very root of hatred, which supports 1 John 4:18: "*He who fears has not been made perfect in love,*" because in this context, when we fear, we cannot love. Proverbs 10:12 says "*Hatred stirs up strife, but love covers all sins.*" Love, therefore, operates outside of the domain of the defense mechanisms that have as their goal maintaining self-approval; love operates in the domain where God's approval reigns.

The response of hatred when a root of fear has taken hold (a stronghold) is not exclusively reserved for people who may bring attention to our weaknesses or faults however; it also applies to subjects. It's amusing, because out of all the subjects that people come into contact with, two stand out as eliciting a strong aversion—math and Jesus. If the subject of math ever comes up in conversation, there is a great likelihood that someone will say, "Ah, I hated math!"

Math is a subject which every child has had to take, and while many children have no problem manipulating numbers, many more struggle. For those who are more conceptual, relying on pictures of reality with which they may form a bridge, math tends to be more difficult. Nonetheless, the feeling of hatred is directly tied to that subject's ability to undermine a child's self-worth or feelings of competence. Due to the mathematical concepts not being understood, a consequential fear of being out of control takes over; in response, the subject of math is derogated as a stupid and worthless pursuit. Invariably, whenever people feel incompetent or inferior, or whenever they cannot obtain an understanding of something themselves, they must put the thing or person that poses the threat down in order to regain a sense of self-worth.

The other subject that causes a similar reaction is the subject of Jesus Christ, and the exact same reasons apply. Lack of understanding always breeds contempt; the narrow-minded, while usually proposing to be more open-minded, strive to gain their sense of equilibrium by derogating the subject or message that has caused them anxiety. The anxiety, in turn, is a simple response to *light* having illuminated a principled truth, like the Word of God, or having illuminated the truth about one's actual hidden feelings that, if accepted, would challenge the person's state of equilibrium. In essence, the defense mechanisms bring to light the subject of resistance in a profound way; defense mechanisms have as their goal the resistance of truth.

By resisting the truth, the defense mechanisms invariably resist the domain of the fruit of the Spirit, which correspond with the Spirit of truth. According to John 16:13: "*When He, the Spirit of truth, has come, He will guide you into all truth.*" Accordingly, we may logically conclude that the end result of the usage and reliance on defense mechanisms is supportive of all falsehood, which counters the Holy Spirit's, and therefore God's, intents and purposes. For as John 8:31-32 says, "*If you abide in My word, you are My disciples indeed. And you shall know the truth, and the truth shall make you free.*" Freedom, therefore, comes from knowledge of the truth. The domain of the defense mechanisms, however, may be rightly considered a house of bondage, for therein is the truth resisted, and where there is no truth, there is no freedom.

Those who are *bound* attain such a state primarily by resisting the fruit of the Spirit and/or by not resisting the anti-fruit. If one resists the fruit of love, they are implicitly opening the door to anti-love which is coupled with the fruit of fear. Without energy from an outside source that may break this bond, as a neutron does in the case of initiating a fission event, it will become stronger and stronger until hatred ensues which leads to backbiting, gossiping, lying, theft, murder, etc. If, on the other hand, one resists the anti-fruit of fear by accepting the fruit of love, which has both understanding and compassion as characteristic properties, fear must flee, which results in a consequent freedom from bondage.

Implicit in this explanation of *bound* or *coupled* resistance is a required expulsion of a stabilizing force, like an electron or a positron during beta decay. The stabilizing force for fear in this case would be anti-love, while the stabilizing force of love would be anti-fear. These stabilizing forces couple with fruit or anti-fruit, and when a change of heart occurs, which corresponds with becoming a new creature in Christ Jesus (see 2 Corinthians 5:17) or the opposite, the stabilizing force must be expelled. This, consequently, allows for a true change to manifest.

While the fruits and anti-fruits are essentially the stabilizing forces, for in order to "*put on love*" the stabilizing anti-fruit of fear must be expelled, energy from somewhere else is often required. The process of "*becoming a new creature in Christ*" accordingly reflects the pattern of beta plus decay, wherein a proton transmutes into a neutron by borrowing energy from a nearby electron (thought) in order to make the transformation. And even though this electron or alternate thought would be close to the heart, which may be counterintuitive due to the fundamental polarity of human beings, we may rightly acknowledge that a plethora of contradicting thoughts are indeed often available for the borrowing.

Therefore, due to the bound or coupled relationship between the fruit and anti-fruit and their stabilizers, in order to eradicate one stabilizing force and admit another, an influx of energy is often required; this source, in turn, appears to assume the form of words. And in the case where one is becoming a new creature in Christ Jesus, this outside *energy* source is most perfectly realized as the Word of God.

Acting, in this context, in one way, as the wind or as a cue ball, the Word of God moves us to go forward in a specified direction. In more precise terms, however, it is a weapon, a "*sword of the Spirit*" (Ephesians 6:17) that binds us to Himself and looses us from our bonds. If we are in bondage to fear, the Word of God looses us from that bond and leads us toward Christ with whom we may become bound instead as the "salt of the earth." Once bound with Him, His thoughts and His Words become the orchestrators of our life's path, which will lead us to freedom and peace. However, mirroring this process, the words of the world will also expel the stabilizing forces of the fruit of the Spirit and will ultimately lead toward the paths of bondage, wherein defense mechanisms become the repositories of response.

Filled with the potential to expel the stabilizing forces of the fruit of the Spirit, the words of the world may be defined as having the opposite characteristics of those things mentioned in Philippians 4:8—things that are true, noble, just, pure, lovely, and of a good report. These words, therefore, could be defined as having the characteristics of things that are false, dishonest, unjust, impure, morally offensive, and of a bad report. Anti-unbelief, therefore, as the stabilizing force of faith, is attained by hearing the Word of God as Romans 10:17 accurately states: "*So then faith comes by hearing, and hearing by the Word of God.*" And anti-love as the stabilizing force of fear, which ultimately leads to hatred, is gained by *hearing* and consequently *believing* words that correspond with any of the characteristics mentioned regarding the words of the world. But here's the crucial point: While the act of understanding certainly may help us to resist the anti-fruit by enabling us to make a conscious choice to resist them as well as all falsehood promoted by defense mechanisms, the true importance of *understanding* rests upon understanding the Word of God.

While understanding may be defined as the power to make sense of something by the application of familiar concepts—according to Proverbs 4:7, *understanding* is considered a secondary gain: *"Wisdom is the principal thing; therefore get wisdom. And in all your getting, get understanding."* According to this Scripture, one must attain wisdom first, and then when wisdom has been attained *understanding* is to be consciously sought. In a very real way then, we may comprehend the notion of understanding as the stabilizing force of wisdom. And according to a dictionary definition, wisdom is the "ability to discern inner qualities and relationships, or insight; or it is the natural ability to understand things that most other people cannot comprehend." First Corinthians, however, defines wisdom in terms of fulfilling God's plan:

For the message of the cross is foolishness to those who are perishing, but to us who are being saved it is the power of God. For it is written: "I will destroy the wisdom of the wise, and bring to nothing the understanding of the prudent."

Where is the wise? Where is the scribe? Where is the disputer of this age? Has not God made foolish the wisdom of this world? For since, in the wisdom of God, the world through wisdom did not know God, it pleased God through the foolishness of the message preached to save those who believe. For Jews request a sign, and Greeks seek after wisdom; but we preach Christ crucified, to the Jews a stumbling block and to the Greeks foolishness, but to those who are called, both Jews and Greeks, Christ the power of God and the wisdom of God (1 Corinthians 1:18-24).

Jesus Christ is therefore the true definition of wisdom, for the acceptance of Jesus opens the door to the Holy Spirit, who is one with God, and therefore God's mind is able to be imparted to all those who do believe.

In essence, therefore, acceptance of Jesus (the Word, light, and wisdom of God) into one's heart provides a domain space where one gains an ability to discern inner qualities and relationships or have insight into things that others may not. Without the wisdom that *is* Jesus, through which one may gain the indwelling of the Holy Spirit of truth, we become as "*he which received seed by the way side*" as spoken of in Matthew 13:19: "*When any one heareth the word of the kingdom, and understandeth it not, then cometh the wicked one, and catcheth [steals] away that which was sown in his heart. This is he which received seed by the way side.*" Provided here is a spiritual example, not only of highly ionizing radiation, whereby electrons are stolen or removed from atoms, but also of understanding as an expelled stabilizer of wisdom, which is Jesus, the Word of God.

Chapter 31
The God Particle

Twenty-four hours a day, researchers try to locate an elusive particle known as the Higgs boson. Utilizing the Large Hadron (particles made up of quarks) Collider, the particles are whipped at high speeds and kept on course with superconducting magnets until they crash into each other with the result of the creation of a menagerie of particles. Researchers are hoping to one day locate the particle on which the Standard Model of particle physics relies, which because of it elusiveness has been dubbed the "God particle."

While the Standard Model isn't considered a complete theory due to its omission of explaining the physics of dark energy, as well as not providing an answer to why there is an imbalanced ratio between matter and anti-matter, it *has* proven to be self-consistent and adequately explains particle behavior. The Higgs boson is vital to the theory, however, because if they could find it, then the existence of a Higgs field would be validated; it is the proposed Higgs field that might explain the symmetry-breaking mechanism by which particles obtain mass. From the standpoint of the laws of conservation as well as annihilation, matter and anti-matter should have cancelled each other out. Obviously in light of the substance of the universe, particles that are responsible for the creation of matter—such as protons, neutrons, and electrons—were obviously in greater abundance than their anti-particles. Referred to as baryogenesis, this question of the asymmetric baryon universe that has enabled life to exist has remained unanswered.

The discovery of the proposed Higgs boson would be equal in importance to the discovery of the photon as the energy carrier for the electromagnetic force. The problem is that while many new particles have been discovered or created in the process of incessant manipulated collisions, no such particle is likely to be found due to the foundational purpose of the scientifically proposed mechanism.

As proposed in *LotK*, the mechanism that needs to be accounted for was designed by God as a pattern after which men might come to Jesus Christ, the Messiah, and fulfill His Word spoken in Ephesians 2:15: *"having abolished in His flesh the enmity, that is, the law of commandments contained in ordinances, so as to create in Himself one new man from the two [Jew and Gentile], thus making peace.*" And the force that this mechanism is a carrier for is mentioned in Ephesians 2:8-10: *"For by grace you have been saved through faith, and that not of yourselves; it is the gift of God, not of works, lest anyone should boast. For we are His workmanship, created in Christ Jesus for good works, which God prepared beforehand that we should walk in them."* Grace would therefore be the mechanism, while faith would constitute the force field.

As God was in the beginning and knows the end—as it says in Isaiah 46:10: "*Declaring the end from the beginning, and from ancient times the things that are not yet done, saying, My counsel shall stand, and I will do all my pleasure*" (KJV) as well as Revelation 1:8 that says, "*I am the Alpha and the Omega, the Beginning and the End*"—it became essential for Him to establish fundamental universal laws that support not only the mechanistic constructs of the natural world, but also would enable Him to tap into these principles when the *appointed time* arrived: "*Because He has appointed a day on which He will judge the world in righteousness by the Man whom He has ordained. He has given assurance of this to all by raising Him from the dead*" (Acts 17:31). Processes that would enable transmutation from one identifying element into another therefore reflect this plan and occur, as we have said, only with left-handed particles during the weak interaction. Considered a symmetry-breaking process, this process that occurs at the particle level during the weak interaction appears to coincide with the symmetry-breaking process that may account for the baryon asymmetrical universe.

In addressing the question of why matter and anti-matter was disproportionately present in the universe in the very beginning, the obvious answer is that something promoted more matter particles to be in existence. If there had been a proportionate amount of matter and anti-matter, according to conservation laws they would have canceled each other out. While CP-symmetry (involving two symmetries—charge conjugation and parity) contends that there should be no change if a particle were to be swapped with its anti-particle, and if left and right, or mirror images, were swapped, in the weak interaction this symmetry is broken. So, under transformation, while left should remain left, right should remain right, negative should remain negative, and positive should remain positive, this assumed conserved pattern is broken. In particular reference to C-symmetry where anti-particles do not remain constant under transformation, this may be understood as validation of the notion that if a man becomes a new creature in Christ Jesus (see 2 Corinthians 5:17) through the pattern of beta decay, all such stabilizing forces that might have identified the man must be expelled.

In light of the dominance of matter over anti-matter, however, it must be acknowledged that Genesis 1:3 says, "*Then God said, 'Let there be light'; and there was light.*" He wasn't recorded as having said, "Let there be light, but let it stop after a million years." And with the consideration that light has been credited in the creation of matter,[6] the command of "*Let there be light*" is a command of continuation, and therefore we see an ever-expanding universe in response to a boundless command.

The combination of this command and the symmetry-breaking that occurred during the foundational outline of systemic processes would appear to correlate with God's plan that would enable both an expanding universe as well as an opportunity for mankind to become renewed. Still, however, the search goes on for substantiated evidence of something that would account for both an expanding universe as well as the baryon asymmetrical universe. Since the beginning of time, men have searched for reasonable answers to the apparently answerable. However, it appears that the answer is much closer to men's hearts than any would suspect. Acts 17:23-30 provides an account of Paul speaking with the men at Athens after he found an altar with the inscription "*To the Unknown God,*" which is simply another way of communicating that God is incapable of being known.

Alluding to God as a planner, whose ultimate plan would involve providing a way for men to actually and substantially *know* God, Paul addressed the men at Athens and gave a call to repentance:

> *To the Unknown God. Whom therefore ye ignorantly worship, him declare I unto you. God that made the world and all things therein, seeing that he is Lord of heaven and earth, dwelleth not in temples made with hands; neither is worshipped with men's hands, as though he needed any thing, seeing he giveth to all life, and breath, and all things; and hath made of one blood all nations of men for to dwell on all the face of the earth, and hath determined the times before appointed, and the bounds of their habitation; that they should seek the Lord, if haply they might feel after him, and find him, though he be not far from every one of us: for in him we live, and move, and have our being; as certain also of your own poets have said, For we are also his offspring. Forasmuch then as we are the offspring of God, we ought not to think that the Godhead is like unto gold, or silver, or stone, graven by art and man's device. And the times of this ignorance God winked at; but now commandeth all men every where to repent (Acts 17:23-30 KJV).*

Paul's call to repentance was a plea to men to open their minds and change their hearts, as the definition of repentance is "to change one's mind" just as he had. Having been a persecutor of Christians, Paul's moment of turning came on the road to Damascus when suddenly there had shone around him a light from

---

6 John A. Gowan, "Section 9 - Symmetry: Noether`s Theorem and Einstein's 'Interval,'" Modified May 2011, accessed January 28, 2014, http://www.johnagowan.org/sect9.html.

heaven: "*Then he fell to the ground, and heard a voice saying to him, 'Saul, Saul, why are you persecuting Me?' And he said, 'Who are You, Lord?' Then the Lord said, 'I am Jesus, whom you are persecuting. It is hard for you to kick against the goads'*" (Acts 9:4-5). At that moment, Paul, whose name had been Saul, turned and chose another direction for his life; he chose to follow the one whom he had spent years trying to defeat. Apparently, Paul had proven to be as a left-handed particle that had been provided grace, or the mechanistic ability to undergo a process of transformation. Interestingly, the response of those around him was no different than it would be today: "*And when they heard of the resurrection of the dead, some mocked, while others said, 'We will hear you again on this matter,'*" meaning, we have nothing to say right now, but we're not done with you. "*However, some men joined him and believed*" (Acts 17:32, 34).

The ability to hear and consequently understand and *turn* from the path on which one is walking is related to the physics concept of chirality, which is a mathematical approach to the concept of handedness, or to the helicity of a particle. Chirality, as applied to chemistry, deals with the symmetry between two objects and specifically refers to an object's relationship to its mirror image. For example, a chiral molecule has a nonsuperposable mirror image, meaning that an object and its mirror image do not exactly correspond. Nonetheless, the two mirror images of a chiral molecule make up a pair; and feeding off of the concept of handedness, they are designated as right- or left-handed. The applicable umbrella concept therefore is symmetry.

As only left-handed particles participate in the weak interaction whereby they undergo the process of nuclear transmutation and assume a new identity, the process of beta decay must be examined. When a neutron transmutes into a proton (beta minus decay), its stabilizer, the electron, is emitted, and due to the neutron's greater mass of 1.29 MeV heavier than a proton, no outside energy is required. However, when the proton transmutes into a neutron, due to its comparably smaller size, energy from the outside is often required and this is accomplished most readily through the process of electron capture; an electron (thought) is *captured* from an inner electron shell near the nucleus.

An analogy may be drawn between this and how a diamagnetic substance, like gold, creates a magnetic field that opposes an externally applied one. With this consideration, it may become clear that the pattern of resistance against a dominant force to which a left-handed particle, like gold, appears to abide is applicable to a person who becomes a "*new creature in Christ Jesus.*" Those who repent, or change course, likewise experience internal resistance to the dominant flow.

A point of interest is that the electron's anti-particle, the positron, is a left-handed particle. While an electron also participates in the weak interaction and is emitted when a neutron transmutes into a proton, it would appear that the weak force decay process that exemplifies God's pattern is more precisely beta-plus decay, where a smaller proton transmutes into a neutron and thereby becomes a *lighter* element.

This hypothesis is reached as it more consistently reflects how things in nature tend to decay into a more stable state. When a proton transmutes into a neutron (beta-plus decay), the atomic number of the element decreases, as opposed to increasing, as with beta-minus decay. The proposal that electrons are synonymous to thoughts, and that a proton needs to capture a thought in order to gain enough energy to make the transformation, is supported by 2 Corinthians 10:5: "*casting down imaginations, and every high thing that exalteth itself against the knowledge of God, and bringing into* ***captivity*** *every thought to the obedience of Christ*" (KJV). Transformation begins with renewing our minds. "*And do not be conformed to this world, but be transformed by the renewing of your mind, that you may prove what is that good and acceptable and perfect will of God*" (Romans 12:2).

While the Higgs boson has been accurately referred to as the "God particle" in that it accounts for the symmetry-breaking processes in question, it is not a particle that can ever be located. Rather, the mechanism for which it is hypothetically held accountable is faith, which according to Hebrews 11:1 is defined: *"Now faith is the substance of things hope for, the evidence of things not seen."* Faith *is* substance! And faith is powered by hope in a particular vision: *"But if we hope for that we see not, then do we with patience wait for it"* (Romans 8:25 KJV). Conversely, *unbelief* is anti-substance, and is powered by anti-hope; however, there's really no such thing as unbelief, as we are all entrained to hold beliefs. *Unbelief* as opposed to faith, however, implies unbelief in God. Accordingly, whatever we are entrained to believe, if it is accompanied by expectation that it will come to pass, has a great probability of coming to pass: "*For the thing which I greatly feared is come upon me, and that which I was afraid of is come unto me*" (Job 3:25 KJV); and "*Therefore I say to you, whatever things you ask when you pray, believe that you receive them, and you will have them*" (Mark 11:24).

We, as human beings, are always bound to a belief of some sort, and the character of that belief will determine our lives. Proverbs 29:18 implies the importance of having in one's mind's eye a goal that, when corresponding with the will of God, leads toward life: "*Where there is no vision, the people perish*" (KJV). And in Habakkuk 2:2-3 it provides an added measure: "*Write the vision and make it plain on tablets, that he may run who reads it. For the vision is yet for an appointed time; but at the end it will speak, and it will not lie. Though it tarries, wait for it; because it will surely come, it will not tarry.*"

When faith and words come into alignment with one another, coupled with the resistant stabilizer of anti-unbelief, there remains no question as to a vision's fulfillment. Therefore, when "*faith which worketh by love*" (Galatians 5:6 KJV) is accompanied by God's Word, the most powerful interaction between light, sound vibrations, and the creative light energy that every man has inherited—because we all have been created in God's image (see Genesis 1:27)—comes to bear. Traveling at a frequency that matches that of pure white light, or pure love, the Word of God mixes with faith, which is "*the substance of things hoped for, the evidence of things not seen*" (Hebrews 11:1). Faith then functions, in this instance, as a substrate or material on which an enzyme acts, or a surface on which an organism grows or is attached. Consequently, when these two things, faith and the Word of God, are combined, the substance of things hoped for becomes evidenced by having been made.

Chapter 32
Nuclear Words

"*For unto us was the gospel preached, as well as unto them: but the word preached did not profit them, not being mixed with faith in them that heard it*" (Hebrews 4:2 KJV). From this Scripture it becomes evident that the Word of God must be received through the force field of faith in order for it to manifest its full potential. As a natural seed requires good ground, sun, and water in order to flourish and yield its fruit, the Word of God requires a prepared heart that has within itself a *receptor seed* that can resonate with subsequent waterings of the Word, which will bring forth the fruit of full-blown faith as Romans 10:17 says, "*Faith comes by hearing, and hearing by the Word of God.*" Accordingly, the combination of faith and the Word of God can manifest with the powerful potential of the compacted energy within a seed.

As pure light within an optical resonant cavity (laser) lined with mirrors bounces light back and forth until a powerful beam of light is generated that can bind that which is broken or loose that which is bound, so too is the Word of God: "*The Spirit of the Lord God is upon me; because the Lord hath anointed me to preach good tidings unto the meek; he hath sent me to bind up the brokenhearted, to proclaim liberty to the captives, and the opening of the prison to them that are bound*" (Isaiah 61:1 KJV). It's no coincidence that the Word of God is referred to as a seed, as a seed has within itself nuclear atomic power; it holds the greatest potential energy, holding within the confines of its parameters the full manifestation of the appointed fruit, flower, or living species. It holds within itself the kinetic energy potential, the gamma radiation potential as well as the free neutron potential that are all products of a fission event, but it converts this energy into matter.

Due to the self-similar nature of a seed, which is patterned after the Word of God, which "*was in the beginning with God and was God*" (John 1:1), it is the earthly manifestation of life-generating power. Accordingly, Jesus—whose alternate identity is the Word of God, as alluded to in 1 John 5:7—has been referred to as the seed of Abraham: "*Now to Abraham and his seed were the promises made. He saith not, And to seeds, as of many; but as of one, And to thy seed, which is Christ*" (Galatians 3:16). Jesus, the Word, holds the self-similar structural ability that all seeds do, whereby those who accept Him by faith as their Lord and Savior have the same potential to become as He is in this world (see 1 John 4:17). We are enabled to adopt His nature and become new creatures in Christ Jesus (see 2 Corinthians 5:17): "*For ye have not received the spirit of bondage again to fear; but ye have received the Spirit of adoption, whereby we cry, Abba, Father*" (Romans 8:15 KJV). Much of this, of course, occurs at the thought/electron/positron level, where either thoughts of one particular nature or another are expelled in a process of renewal.

The nuclear power that is stored in a seed is the same power that is not only found within the Word of God, but is stored within all words as well. And words—like the Word of God, or like any seed—produce after their own nature. If words correspond to the fruit of the Spirit, the product or fruit will have a similar nature, and the same *rule* applies when words correspond to the anti-fruit of the Spirit: "*every good tree bringeth forth good fruit; but a corrupt tree bringeth forth evil fruit*" (Matthew 7:17 KJV). "*And the tongue is a fire, a world of iniquity: so is the tongue among our members, that it defileth the whole body, and setteth on fire the course of nature; and it is set on fire of hell*" (James 3:6 KJV).

When anti-fruit is in dominion, the light-absorbing dark energy has the potential of halting light's creative power, which is rooted in love. As a consequence, the stabilizing force (like an electron or a positron, or anti-unbelief for faith or anti-fear for love, for example) will be emitted, which opens a *door* to new signal frequencies to be received. This admittance, in turn, brings about changes in our thought (electron) patterns, which is accompanied by a corresponding change in words and/or behavior. While under many circumstances actions do, in fact, speak louder than words, the *seed* potential of words is perennial; wounding words continually break through the soil of hearts—our own as well as others'. They are as weeds, requiring being dug up by the roots and discarded and replaced by new seed.

Functioning like neutrons in a nuclear reactor—as instigators and sustainers of nuclear-like chain reactions—words are both expelled from a heart as well as are absorbed by a heart, for "*out of the abundance of the heart the mouth speaks*" (Matthew 12:34). In the natural realm, when a free neutron, which is unstable, is absorbed by a large atomic nucleus that is able to undergo fission, such as uranium-235 or plutonium-239, the heavy nucleus splits into two or more lighter nuclei; consequently, kinetic energy, gamma radiation, and free neutrons are released, which would imply that they already existed—at least, in a potential form. Similarly when words, which operate as waves of a particular light frequency, are absorbed into the heart of man, their frequencies have an amplitude and velocity that has been stored in a form of potential energy.

When words are spoken, it is equivalent to the process of nuclear fission in terms of power. In a nuclear reactor, when neutrons are absorbed by a heavy nucleus through collisions, they merge to form a heavier nucleus. Since neutrons, like words that are essentially neutral in and of themselves, lack charge, they are easily absorbed by colliding nuclei as they experience no initial electrostatic repulsion. During the process of fission, more free neutrons (carriers of the intrinsic power of words) are emitted, which then collide with other nuclei, by which they are absorbed and cause the same process to continue; as an example, the pattern of strife—or its opposite, praise, carried out through words in a social or corporate setting—is thus equal to the process of a nuclear chain reaction.

Automatic and manual systems that control the rate at which neutrons are absorbed by colliding nuclei are established in nuclear reactors. Similarly, we have been provided the Word of God with which we may stop the devastating effects of words destroying our own as well as others' lives. The devastation caused by fire is unparalleled; everything is utterly destroyed: "*And the tongue is a fire, a world of iniquity: so is the tongue among our members, that it defileth the whole body, and setteth on fire the course of nature; and it is set on fire of hell*" (James 3:6 KJV). And as nuclear fission, caused by neutron absorption, is converted to thermal energy by a nuclear reactor, so are words indicative of our own spiritual temperature, whether hot, cold, or lukewarm: "*I know thy works, that thou art neither cold nor hot: I would thou wert cold or hot. So then because thou art lukewarm, and neither cold nor hot, I will spue thee out of my mouth*" (Revelation 3:16 KJV).

Clarifying that "works" relates to the words one lives by, Revelation 3:8 pronounces: "*I know your works. See, I have set before you an open door, and no one can shut it; for you have a little strength, have kept My word, and have not denied My name.*" The words of a particular frequency that we absorb are the ones that we emit, just like an electron emits a photon of a similar frequency to the one that it absorbed. The frequencies we agree with, having *receptor seeds* within our hearts that have already taken root, will most readily be absorbed. Many words, however, that come from left field so to speak will also be readily absorbed; this occurs because similar to neutral neutrons that experience no electrostatic repulsion, words intrinsically experience little resistance; instead it takes work to resist words entering and consequently getting a foothold in our hearts. Highlighting our vulnerability, 1 Peter 5:8-9 declares a truth: "*the devil, as a roaring lion, walketh about, seeking whom he may devour: whom resist steadfast in the faith*" (KJV). The greatest way he can accomplish this is by lowering our immunity to the absorption of destructive words, so that our resistances are basically inoperable. He is lord of the anti-fruit and desires to deceive us into believing that words have no power; by believing him, we become disempowered by continuing to operate accordingly.

The control systems in a nuclear reactor vary. While a neutron moderator is a medium that slows fast neutrons down so they will become thermal neutrons, which can sustain nuclear chain reactions, neutron poisons absorb neutrons so they are unavailable for a nuclear reaction process. While a moderator slows neutrons, a neutron poison absorbs them, basically making them ineffectual. Interestingly enough, graphite is a neutron moderator as is water and heavy water (deuterium) and may be understood as affecting the process by which the Word of God is absorbed and reemitted. With regard to graphite, this is a conclusion drawn from understanding that conductance from frequencies beyond the planar level are restricted. With consideration of the graphite principle mentioned earlier, a parallel may be drawn: In the event that words are of a fruitful nature, as through prayer or praise, the introduction of one who is bound to this earthly level can slow down the power of the Word and Spirit; it can't kill it, but it can dampen the effectiveness. In terms of water and heavy water, not only does water effectively absorb neutrons (word carriers), but the similar size of neutrons and the hydrogen atoms within the water allows for efficient displacement. These moderating mechanisms thus deter nuclear chain reactions, whether good or bad.

With regard to a neutron poison, both the Word of God as well as words of an anti-fruitful nature can be considered applicable. The Word of God as a neutron poison (absorber) is applicable to those times when the anti-fruit and thus anti-fruitful words are self-replicating. In this context, the Word of God has the capacity of absorbing or swallowing up the negatively generated nuclear-like power, with the end result of diminished reactivity. And in the same way, a neutron absorber may also be understood as any word that slows the Word of God's absorption, i.e., words of doubt, unbelief, and fear. While neutron poisons are regarded as mostly undesirable in the context of a nuclear reactor's general purpose, to set off a nuclear chain reaction, the essential neutrality of words demands emphasis on context. As neutrons are essentially neutral but are at the same time the key to nuclear power, understanding words as equivalent in function and form lends support to their reliance on context to be fully understood. However, in the sense that neutron poisons can help lower the initial high reactivity that occurs when fuel is loaded into the reactor, we may understand "*the sword of the Spirit, which is the Word of God*" (Ephesians 6:17) as performing a similar role of absorbing the enemy's strength.

Under the condition that a neutron poison reaches the reactor core where its effect goes beyond moderating the power, however, the result can be a complete stalling of the chain reaction. This example provides an analogy of what occurs when someone who has been absorbing and reemitting thoughts and words of one particular character, emulating a chain reaction, suddenly is moved deeply by thoughts or words that reach into the heart. When this occurs, with different feelings having been struck, there is an impaired response from the mechanism that would automatically reject the intrusion. The end result is a stalling of the system, which may be seen as equivalent to the moment of an epiphany, where introspection and consideration take the place of responding automatically according to the prior patterned response. Of course, this example would occur when the intrusive thought or word that represents a neutron poison corresponds with truth and/or the fruit of the Spirit.

The heart, therefore, as a representative reactor core is that which ultimately determines whether or not a nuclear-like chain reaction will occur. Similar to how the heart ultimately determines how the good seed of the Word is heard and therefore acted upon, the reactor core is where the process of fission or *loosing* occurs. Effectively, this may be understood as the place where bondages are essentially broken—in the place where heart and mind connect and affect *breakthroughs*. Loosing then is not only a key to the kingdom of heaven, which is but one function of the Word of God, but in this context it is the process whereby one's words as well as actions are sustained.

Paralleling the documentable yield that planted seed might bring forth, nuclear reactor systems utilize an instrument, which basically categorizes the *fruitfulness* of a nuclear chain reaction. In the same way that good seed when it is sown on good ground may yield different percentages of return according to Matthew 13:23, the nuclear reactor system utilizes something called an effective neutron multiplication factor, which basically categorizes the fruitfulness of a nuclear chain reaction. Referred to as subcritical, critical, and supercritical, these levels are of equivalent comparative value to thirty, sixty, and hundredfold returns; and while the levels of thirty, sixty, subcritical, and critical are not optimum for both processes, supercritical and hundredfold both exemplify an abundant return for the input into the *systems*.

With words conceptually behaving like neutrons, both initiating as well as sustaining nuclear-like chain reactions, which consequently proceeds out of the heart according to Matthew 12:34: *"out of the abundance of the heart the mouth speaks,"* we may understand the process of fission in a new light. When the Word of God, which is the laser-like "*sword of the Spirit*" is absorbed, like a neutron's absorption causes a heavy nucleus to split, "*the word of God is living and powerful, and sharper than any two-edged sword, piercing even to the division of soul and spirit, and of joints and marrow, and is a discerner of the thoughts and intents of the heart*" (Hebrews 4:12). In other words, the Word also causes a split or a division of the nucleus/core/heart to occur. In addition, with regard to the light that is emitted during the process of fission, it has been established that people emit frequencies of light through the eyes, and these frequencies can resonate with the fruit of the Spirit or with its anti-fruit. The light that is released with the free neutrons, therefore, will correspond to the light of love, which is God, or with the dark light of fear and the father of it.

Words, however, and most precisely the Word of God, have power, and as Hebrews 4:12 declares, they can *loosen,* through the process of fission, the strongholds in a heart. Not only that, words that escape through the mouth become coupled with these light frequencies, so as neutrons are released by a splitting nucleus with a simultaneous release of radioactive light, words are released out of the mouth and a corresponding emission of a light frequency also occurs. Now we return to the notion of interference, for as waves of light are emitted, they intercept other light emissions and can either constructively or destructively interfere with them. Nonetheless, when more and more Word is emitted through the process of fission and consequently absorbed by more and more nuclei (hearts), not only does faith come according to Romans 10:17, but it becomes the foundation for revival.

In the case that the Word of God is involved in a nuclear-like chain reaction, such as what would occur in a revival, which would invariably be pleasing to God, a neutron poison would absorb or steal the Word from being able to be *used.* In this situation, obviously, the neutron poison would be equivalent to the Devil who steals the Word out of our hearts, according to Matthew 13:19: "*When anyone hears the word of the kingdom, and does not understand it, then the wicked one comes and snatches away what was sown in his heart. This is he who received seed by the wayside.*" When the Word of God, however, takes on the role of a neutron poison and slows or stalls an unfruitful chain reaction, the situation mirrors Saul on the road to Damascus.

Chapter 33
Covenants

I went to the beach a while back in search of mussels, and while many had small passengers stuck on their back, I found one that was strongly adhered to a rock. Requiring a knife to loosen the hold, I was struck by the strength of the connection; it was an "Aha!" moment. The strong bond between this mussel and this rock perfectly illustrated the covenant relationship that God has made with His children. If we don't sever the connection ourselves—as with a knife, or willful disobedience without repentance—the covenant relationship between God and His children, who are heirs together with Jesus, is as strong as the relationship that God the Father and Jesus the Word have; they are one, just as the mussel and the rock had become one.

Associated with the nuclear strong force, the topic of covenants brings to light a connection with the fulfillment of God's gift of human potential. In the same way that the mussel traveled on whatever path the rock traveled or was planted in whatever ground the rock was planted, so too is it with those who are in a covenant relationship with God. As the mussel's adhesion to the rock parallels a strong force interaction, what becomes evident is that the mussel's potential was no longer determined by its own strength, but rather by the rock's strength. If the rock was carried by a wave and landed high on a cliff, there the mussel would be; it would have gained potential energy, albeit in a mechanical sense, but nonetheless its ability to affect *reality* would have changed. Accordingly, those who have entered into a covenant relationship with God and rely on His potential to define their own true potential will see impossibilities become probabilities.

As energy carriers for the strong force, gluons, which are responsible for the strong bond between nucleons, may be understood as analogous to the binding agents of a covenant. The energy that gluons carry, referred to as *color charge*—so named because of a similarity of structure with the primary colors, yellow, blue, and red—is the specific force within the gluons, which passes between the quarks that make up protons and neutrons. The theoretical underpinnings of this concept may best be understood if a color wheel on which these three colors are placed is considered to be representative of the three quarks within both a proton and neutron. The concept of color charge is then based upon the proposal that when a gluon passes between these quarks, its charge, which is affected by the quarks during the interaction, in turn affects the subsequent quark; the process, therefore, parallels the way colors affect other colors when mixed. Nonetheless, the continuously changing charge, like pollen picked up and deposited by a bee, performs the function of glue and consequently binds the quarks within protons and neutrons inside of a nucleus. In this way, a gluon functions as a binding agent or as the heavenly patterned mechanism attributed to the power of an oath, or a covenant; the implication of this, of course, is that gluons are sent forth by words, as words are the binding agents of an oath or covenant.

In accordance with the process involved in fusion, the forming of a covenant and a strong bond involves a loss of heat energy, which signifies a consequent loss of potential. However, if light is involved in the creation of matter, then light must be stored somewhere in all creation. Thereby, even if heat dissipates, the light (electromagnetic radiation) potential remains in a form of latent potential energy.

While the concept of potential energy is generally associated with the physical restoring of forces, such as the restoration of a spring after having been stretched, *LotK* extends its application to include all latent heat or light potential. According to phase transitional principles, the loss of heat energy describes a process where potential decreases, such as when vapor transitions into a solid. However, in nuclei that can undergo fission, the loss of light energy that occurs when the nuclei splits appears to have been stored in a form of potential light energy that manifests only when a neutron (word) is absorbed by the nucleus (heart). Therefore, what becomes evident is that this stored light potential is an integral part of what occurs during the binding process, or during the process of establishing a covenant the way God did with Noah, Abraham, Jesus, and by extension, those who love Him.

The other side of the process, however, cannot be ignored. While loss of heat energy and a gain of potential light energy occur during fusion, a gain of heat energy and a loss of potential light energy occur during fission. While the gain of potential light energy during fusion is lost during fission in that that light energy is emitted, what intrinsically occurs is a manifestation of light's kinetic energy, where light is able to do its work, or in other words, create matter. Essentially, when God said, "*Let there be light,*" He was *loosing* light from its bondage to the potential realm. And as both faith and fear belong in the potential realm, loosing (fission) can have beneficial as well as disastrous consequences. While faith loosens reflective white light from the potential realm, fear loosens absorbing dark light from the potential realm. However, while fission loosens light from a realm of bondage entered into during fusion, due to the inherent neutrality of matter, fission and fusion, like all other polarities, must be valued contextually.

Along with being considered instigators and sustainers of a nuclear chain reaction, as are neutrons, words may also be considered as binding agents, like gluons, that establish oaths, strong bonds, and covenants. During the fusing (entering into a covenant) process, the release of words may be considered equivalent to the release of heat energy. However, when spoken words are released and are mixed with faith in their truth and value, light energy becomes stored as potential energy. If the covenant is obeyed, this storage of potential energy remains bursting with potential for the participants; however, if words of unbelief, doubt, and fear are absorbed into the heart that makes one doubt the strength of the covenant, bond, or oath, then fission can occur, severing the union.

Accordingly, with fission comes the expulsion of more words (free neutrons), which can likewise be absorbed by other nuclei (hearts) and thereby a social, usually chain reaction ensues and often carries explosive power—i.e., gossip, backbiting, or jealousy. In the same way, however, the Word of God can be used to affect a breakthrough which entails severing the bonds of a stronghold, which can result in an explosively powerful chain reaction of praise and thanksgiving. According to *LotK*, therefore, the absorption of the words of the world or the absorption of the Word of God will determine the bonds that we make and the covenants into which we enter.

As we understand it, destructive potential resides within radioactive elements, and by extension, the radioactive release in a fission event accurately illustrates the dynamic that occurs during many break-ups, divorces, or splits of any sort. Due to a commitment-forming bond and the consequent diffusing of the elements involved, which lowers the total energy, the energy that breaks the bond (negative binding energy) has to be stronger. And the product of such a break often mimics the gamma radiation that is released during nuclear fission.

Psychologically, when a bond is broken, feelings mount—of rejection, fear, failure, and a need to protect oneself. As with reaction formation, where people express the opposite of what they feel, this process accounts for the tendency of ex-partners to express toward one another an overcompensating degree of contempt. Due to the tendency for an intimate bond between people to be characterized according to the operation of the fruit of the Spirit, when there is dissolution of that bond, a stronger energy like negative binding energy needs to enter in order to overcome the positive binding energy. However, as in one product of a fission event, when covenants are broken, energy of an explosive nature, like gamma radiation, is often produced.

Along similar lines yet less extreme, the breaking of such a bond may also be understood according to wave dynamics and interference. For example, when two waves that have become bound through constructive interference are suddenly out of phase with one another, not only must one assume that destructive interference occurred to break the harmony, but also that destructive interference is utilized at the heart level to deal with the break. In order to eradicate feelings of love, represented by a waveform with a positive amplitude (crest), a waveform of the same value but with a negative amplitude (trough) effectively brings it to a flat line level if thrust in its direction. This often occurs by turning one's feelings of love into hatred, respect into disrespect, or simple caring into not caring at all. An effective defense mechanism that keeps the pain of rejection and loss at a minimum, reaction formation, like all defense mechanisms, keeps the truth at bay.

Nonetheless, by aligning one's thoughts with that which will destroy memories, particularly of the old waveform, the old waveform loses strength and allows people to move on with their lives. Regardless of the feelings connected with a strong bond, however, what appears essential for self-preservation is the destruction of the memory of the original waveform. Regardless of the analogy, however, when covenants are broken, the consequences can range from gamma radiation-like explosions to flatline waveforms that may aid in restoring a sense of emotional equilibrium.

When we are bound to others, due to the fact that a primary human need is to belong, our social need often takes precedence over any value-driven directive. However, if we are bound to someone who shares our values, our conscience is rarely divided, or rather polarized. Similarly, we can loosen ourselves from people who pose a threat to our values, or we can loosen ourselves from those whose *light* exposes the truth within us. It's all relative. What is clear, however, is that in terms of binding and loosing, the more beneficial end results will be coupled with the fruit of the Spirit, whereas the less desirable will be coupled with its anti-fruit. In effect, while *loosing* oneself from a covenant that is coupled with the fruit of the Spirit may not be the wisest choice, *loosing* oneself from a covenant coupled with the anti-fruit is certainly a valuable prospect.

Verification of the choice between that to which we bind or loose ourselves appears in Isaiah:

> *Because you have said, "We have made a covenant with death, and with Sheol we are in agreement. When the overflowing scourge passes through, it will not come to us, for we have made lies our refuge, and under falsehood we have hidden ourselves." Therefore thus says the Lord God: "Behold, I lay in Zion a stone for a foundation, a tried stone, a precious cornerstone, a sure foundation; whoever believes will not act hastily. Also I will make justice the measuring line, and righteousness the plummet; the hail will sweep away the refuge of lies, and the waters will overflow the hiding place. Your covenant with death will be annulled, and your agreement with Sheol will not stand; when the overflowing scourge passes through, Then you will be trampled down by it (Isaiah 28:15-18).*

In short, binding oneself to a covenant with death is a not a smart bet, while binding oneself to a covenant with God and His Son, Jesus Christ, brings righteousness and life.

In support of the proposal that the domain of the anti-fruit harbors things that lead to death rather than to life, 2 Corinthians 7:10 makes a proclamation: "*For godly sorrow produces repentance leading to salvation, not to be regretted; but the sorrow of the world produces death.*" As joy is a fruit of the Spirit, mentioned in Galatians 5:22, we know that its opposite, sorrow, is an anti-fruit. In light of this Scripture, we know that the end of this anti-fruit is death. Death opposes creation as well as all creativity, and due to this opposition a covenant with death is a bond with decay. While a covenant with God provides unexpected "suddenlys" to emerge, a covenant with death traps and confines, as if in a well. Light creates; darkness annihilates. Only a covenant with the light provides creative hope as well as a way to fulfill one's true potential.

In conclusion, in relation to the process of fusion that reflects the physical process of entering into and remaining in a covenant relationship, the proposed storage of potential light energy is commensurate to the gamma radiation released during fission. Existing in a dormant form *within* the covenant relationship, this light energy is stored as potential energy. A covenant with God will therefore be infused with all of His *light* power, ready to be released when His Word's power enters the heart. Conversely, a covenant with death will be infused with all of Satan's dark power, ready to be released when his words enter the heart instead. Perhaps analogous to a pressure cooker, tea kettle, volcano, or seed, at some critical point all of the potential energy, whether dark or light, will be made manifest. The fact that the energy remains often undetectable has little bearing on the reality of its potential to burst forth with powerful potential. Accordingly, we may understand our own potential's manifestation as adhering to this pattern. The stored potential depends upon the covenant, whether of life or death, and the release of the potential depends upon the character of the words absorbed.

Within the domain of covenants, the subject of choice has vital consequences. While God's covenant with the universal entities which operate according to gravitational laws appears unbreakable—i.e., with the stars, planets, seasons, etc.—due to man having been provided free will, covenants are determined by obedience. For instance, God entered into covenant with Noah because of Noah's obedience to build an ark and offer praiseworthy sacrifices on an altar to God. In response to this obedience, God said to Noah: "*I will never again curse the ground for man's sake, although the imagination of man's heart is evil from his youth; nor will I again destroy every living thing as I have done. While the earth remains, seedtime and harvest, cold and heat, winter and summer, and day and night shall not cease*" (Genesis 8:21-22). God coupled His covenant with

Noah to the already established covenant that He had originally entered into with the elements of the universal domain.

In Genesis, God lays out the specifics of the covenant:

> *"And as for Me, behold, I establish My covenant with you and with your descendants after you.... Thus I establish My covenant with you: Never again shall all flesh be cut off by the waters of the flood; never again shall there be a flood to destroy the earth." And God said: "This is the sign of the covenant which I make between Me and you, and every living creature that is with you, for perpetual generations: I set My rainbow in the cloud, and it shall be for the sign of the covenant between Me and the earth" (Genesis 9:9, 11-13).*

In confirmation of the purpose for God's covenants with man, Ezekiel 1:28 pronounces, *"Like the appearance of a rainbow in a cloud on a rainy day, so was the appearance of the brightness all around it. This was the appearance of the likeness of the glory of the Lord. So when I saw it, I fell on my face, and I heard a voice of One speaking."* The strong bond of a covenant with God signifies, above all, His desire to speak with us and be in an intimate relationship with us. As it says in Genesis 6, "*Noah found grace in the eyes of the Lord...Noah was a just man, perfect in his generations. Noah walked with God*" (Genesis 6:8-9). *"For the eyes of the Lord run to and fro throughout the whole earth, to show Himself strong in behalf of those whose heart is loyal to Him"* (2 Chronicles 16:9).

Noah, however, was not the only man with a strong relationship with God. Abram was also such a man. In light of this relationship, God chose Abraham to fulfill His plan on earth.

> *Now the Lord had said to Abram: "Get out of your country, from your family and from your father's house, to a land that I will show you. I will make you a great nation; I will bless you and make your name great; and you shall be a blessing" (Genesis 12:1-2).*

> *When Abram was ninety-nine years old, the Lord appeared to Abram and said to him, "I am Almighty God; walk before Me and be blameless. And I will make My covenant between Me and you, and will multiply you exceedingly." Then Abram fell on his face, and God talked with him, saying: "As for Me, behold, My covenant is with you, and you shall be a father of many nations. No longer shall your name be called Abram, but your name shall be Abraham; for I have made you a father of many nations. I will make you exceedingly fruitful; and I will make nations of you, and kings shall come from you. And I will establish My covenant between Me and you and your descendants after you in their*

generations, for an everlasting covenant, to be God to you and your descendants after you" (Genesis 17:1-7).

Those who accept Jesus Christ by faith—who accept Him as their personal Lord and Savior—are accounted as Abraham's seed: "*Therefore it is of faith, that it might be by grace; to the end the promise might be sure to all the seed; not to that only which is of the law, but to that also which is of the faith of Abraham; who is the father of us all, (as it is written, I have made thee a father of many nations,) before him whom he believed, even God, who quickeneth the dead, and calleth those things which be not as though they were*" (Romans 4:16-17 KJV).

While faith is often referred to as a belief that is held, *working faith* requires obedience, among other things.

By faith Noah, being divinely warned of things not yet seen, moved with godly fear, prepared an ark for the saving of his household, by which he condemned the world and became heir of the righteousness which is according to faith. By faith Abraham obeyed when he was called to go out to the place which he would receive as an inheritance. And he went out, not knowing where he was going. By faith he dwelt in the land of promise as in a foreign country, dwelling in tents with Isaac and Jacob, the heirs with him of the same promise (Hebrews 11:7-9).

By faith Abraham, when he was tested, offered up Isaac, and he who had received the promises offered up his only begotten son, of whom it was said, "In Isaac your seed shall be called" (Hebrews 11:17-18).

Faith in God and obedience to the Word of God (Jesus) brings salvation and a consequent rest: "*Come unto me, all ye that labour and are heavy laden, and I will give you rest. Take my yoke upon you, and learn of me; for I am meek and lowly [humble] in heart: and ye shall find rest unto your souls*" (Matthew 11:28-29 KJV). Obedience to His Word here is simply to "*Come.*"

Abraham had believed God for a son, regardless of his and his wife, Sarah's, old age (see Romans 4:3). He was told that he would be the father of many nations and then was put to the test, being asked to sacrifice Isaac, his only son for whom he had believed. Despite the obvious angst that such a request evoked, Abraham was duly obedient and proceeded to honor God. And it paid off, for at the last minute, "*The angel of the Lord called unto him out of heaven, and said, Abraham, Abraham: and he said, Here am I. And he said, Lay not thine hand upon the lad, neither do thou any thing unto him: for now I know that thou fearest God, seeing thou hast not withheld thy son, thine only son from me*" (Genesis 22:11-12). And a ram caught in the thicket was provided for the sacrifice instead. Abraham trusted God and didn't rely on his human reasoning.

While most of us would submit to the Serpent's beguiling question as in the garden of Eden, "*Yea, hath God said...?*" (Genesis 3:1), Abraham didn't question the command; he followed through on His faith in God with works of obedience to what God had commanded: "*Yea, a man may say, Thou hast faith, and I have works: shew me thy faith without thy works, and I will shew thee my faith by my works*" (James 2:18 KJV). "*Was not Abraham our father justified by works, when he had offered Isaac his son upon the altar? Seest thou how faith wrought with his works, and by works was faith made perfect? And the scripture was fulfilled which saith, Abraham believed God, and it was imputed unto him for righteousness: and he was called the Friend of God*" (James 2:21-23 KJV).

We are Abraham's seed if we have accepted Jesus into our lives, and the promises that were bestowed on Abraham likewise belong to us: "*Now to Abraham and his Seed were the promises made. He does not say, 'And to seeds,' as of many, but as of one, 'And to your Seed,' who is Christ. And this I say, that the law, which was four hundred and thirty years later, cannot annul the covenant that was confirmed before by God in Christ, that it should make the promise of no effect. For if the inheritance is of the law, it is no longer of promise; but God gave it to Abraham by promise*" (Galatians 3:16-18). The strong bond of a covenant presupposes truth, integrity, and an obedient heart as backbones that support the words, intentions, and faithfulness of those who enter it.

The negative binding, or loosing energy that breaks the strong bond of a covenant is therefore based in disobedience. While loosing can very well signify freedom from a bond, similar to how loosing heat energy from liquid or vapor initiates a phase with less potential, so too does disobedience to the Word of God.

The interesting thing is that God wants His children to be free: "*Thus says the Lord, the God of Israel: 'I made a covenant with your fathers in the day that I brought them out of the land of Egypt, out of the house of bondage, saying, "At the end of seven years let every man set free his Hebrew brother, who has been sold to him; and when he has served you six years, you shall let him go free from you." But your fathers did not obey Me nor incline their ear'*" (Jeremiah 34:13-14). God wanted to provide Israel, His chosen people, as well as those who are Abraham's seed with not only freedom but prosperity. However, God realized that He was dealing with children who were basically disobedient, like electrons each pointing in their own direction. But He had made a covenant with their fathers, which was based on the Ten Commandments given to Moses that had the strength of the strong force as a backing; as God is not a liar (see Numbers 23:19), He had to follow through with the conditions of breaking the agreement, which like a fission event is often explosive.

Jeremiah 11:3 proclaims the result of disobedience: "*Cursed is the man who does not obey the words of this covenant.*" Jeremiah was commanded by the Lord to proclaim all the words of God's covenant in the cities of Judah and in the streets of Jerusalem saying, *"Hear the words of this covenant and do them. For I earnestly exhorted your fathers in the day I brought them up out of the land of Egypt, until this day, rising early and exhorting, saying, 'Obey My voice.' Yet they did not obey or incline their ear, but everyone followed the dictates of his evil heart; therefore I will bring upon them all the words of this covenant, which I commanded them to do, but which they have not done*" (Jeremiah 11:6-8). Deuteronomy 28:16-68 also expounds upon these curses fully, which, in the same vein as anti-fruits, are directly opposed to the blessings mentioned in verses 2 to 15 of the same chapter.

The old covenant promised life for obedience and death for disobedience, and because of Adam's initial disobedience to God's command in Genesis 2:17, *"But of the tree of the knowledge of good and evil you shall not eat, for in the day that you eat of it you shall surely die*," humanity was dealt a blow. While an instant physical death was not implied, a spiritual separation was, and spiritual death came upon us all as a consequence of this disobedience.

The initial disobedience to the Word of God resulted in the severing of the initial covenant with God, which most readily would epitomize a covenant of grace, for God had given only one command: "*And the Lord God commanded the man, saying, 'Of every tree of the garden you may freely eat; but of the tree of the knowledge of good and evil you shall not eat, for in the day that you eat of it you shall surely die'*" (Genesis 2:16-17). Before eating of the tree in disobedience, Adam and Eve knew no sin; they were free to be in charge of their world with no consequences. Consciousness of sin—separation from the grace that they had been born into—came only after the Serpent had deceived Adam and Eve into questioning God's Word, as well as His intentions.

"*Yea, has God said?*" were the infamous words that were spoken that loosed Adam's and Eve's thoughts from being bound to God's thoughts. They had been the original "*salt of the earth.*" Being incomplete by themselves, they had joined in an ionic bond with the Word of God and were at rest until the Serpent's dissolution of their *covenant of salt* by steering their minds and hearts away from God's Word. That they had been joined with God as friends is confirmed by the simple suggestion that they were hearing His voice; they were talking and walking with Him. Consequently, it may be assumed that they experienced what is now once again available through the Holy Spirit, an ability to be one with God.

The fact that we who have accepted Jesus into our hearts are referred to as the "*salt of the earth*" in Matthew 5:13 is not without significance. As in the bond of a covenant with God, the creation of salt provides great potential. Although a salt is a bound mixed entity, its mixed state signifies a bond with Jesus Christ and stores within itself electrolytic potential, which, serves to convert chemical energy into electrical energy and electrical energy into chemical energy. With the understanding that a chemical is defined by its ability to cause a reaction in a substance, with faith as a substance that comes by hearing the Word of God, we may understand how faith in the Word of God serves as a chemical that is converted into electrical power when the one who is speaking the Word in faith is bound with Jesus and is the salt of the earth.

These energies are stored in the bound salt as potential, but when the union is dissolved and split, like what occurs during nuclear fission, the potential of the stored energy is realized. In this case, when a salt is dissolved in water, it becomes a representation of what occurs in the process of baptism; the results are electrolytes or ions that acquire the capacity to conduct electricity. When a person is connected to Jesus, who behaves in this context as a negative ion (anion), they become the *salt of the earth* with the ability to conduct His love and His Word when immersed in the Holy Spirit (water).

In explanation of this reasoning, in John 3:30 John the Baptist proclaims with reference to Jesus: *"He must increase, but I must decrease."* Under the condition that representative cations and anions must be joined in order for a salt to be created, the question of which one increases may arise. With an anion (negative ion) having more electrons (thoughts) and a cation having too few, it would appear that it is the anion that needs to decrease the number of electrons that it has and offer those to the cation. According to this reasoning then, *we* may very well represent the positively charged cations that are in need of the excess electrons that are creating the anion's negative charge. In this scenario, we would decrease in the humble sense, as it says in 1 Peter 5:6-7: *"humble yourselves under the mighty hand of God, that He may exalt you in due time, casting all your care upon Him, for He cares for you."*

With humility we would accept His thoughts as our own, and in the process our thoughts and cares would become one with His. Thereby the statement made in 1 Corinthians 2:16 would make logical sense: *"For 'who has known the mind of the Lord that he may instruct Him?' But we have the mind of Christ."* If we bind our thoughts with His thoughts, then we are in the position of having acquired His mind.

Evidently, when we don't allow His thoughts to become ours, it is akin to breaking a covenant. What occurs when a salt bond is broken is a dulling of electrolytic potential, or in other words the loss of the salt's savor as referenced in Matthew 5:13: *"but if the salt have lost his savour, wherewith shall it be salted? it is thenceforth good for nothing, but to be cast out, and to be trodden under foot of men"* (KJV).

This dulling of electrolytic potential that parallels "*salt having lost its savor*" may also be understood as occurring in those who may be considered "*lukewarm.*" Revelation 3:15-16 mentions this state: *"I know thy works, that thou art neither cold nor hot: I would thou wert cold or hot. So then because thou art lukewarm, and neither cold nor hot, I will spue thee out of my mouth"* (KJV). While the temperature conditions of "hot" and "cold" appear to be justified by God, the condition of being "lukewarm" is not. Why? As we review the global phenomena that occur at each of these temperature conditions, it becomes apparent that the supernatural processes like superconductivity and superfluidity occur at extremely cold temperatures, and the parameter of heat leads molecules to transition into vapor, which is then able to readily escape earthly or solid bonds. "Lukewarm" provides for nothing but stagnation. If we apply these conditions to people, we may readily understand these terms' applicability to human "heart conditions," and how God may view the children of men according to these terms.

In addition to these parallels, however, the conditions of hot, cold, and lukewarm may also reference neutron temperatures, which define their energy ranges. While hot neutrons have an energy of about 0.2 eV, thermal neutrons have an energy of about 0.025 eV, and cold neutrons have an energy from 5×10^{-5} eV to 0.025 eV. One important consequence of these temperatures is that thermal neutrons are more easily absorbed due to their lukewarm speed, which makes their capture most accessible to radioactive fissile elements. While in one sense, these temperature speeds of neutrons account for why words spoken to those with an incendiary nature are often captured and reemitted with a vengeance, in another sense, this concept applies readily to social persuasion; when a sender of a message (hot or cold neutron) conveys only one side of the argument, there will be resistance from those who are educated. If both sides of the argument are posed, however, while the uneducated may not know how to respond, the perceived neutrality (lukewarm neutron) of the sender is met with less resistance by those who rightly divide the truth from falsehood.

In addition, hot, cold, and thermal neutron radiation commonly is used in neutron scattering experiments where the radiation is used for the analysis of condensed matter. Therefore, by extension, one could say that with neutrons as word carriers, the character of the words spoken may aptly detect the *heart-level* properties of a man. In turn, these properties of the heart will, by and large, determine all bonds and covenants that are entered into.

Jesus came to earth to provide a new covenant to those who would believe, but He also came to fulfill the law, or the first commandment:

> *Think not that I am come to destroy the law, or the prophets: I am not come to destroy, but to fulfil. For verily I say unto you, Till heaven and earth pass, one jot or one tittle shall in no wise pass from the law, till all be fulfilled. Whosoever therefore shall break one of these least*

> *commandments, and shall teach men so, he shall be called the least in the kingdom of heaven: but whosoever shall do and teach them, the same shall be called great in the kingdom of heaven. For I say unto you, That except your righteousness shall exceed the righteousness of the scribes and Pharisees, ye shall in no case enter into the kingdom of heaven (Matthew 5:17-20 KJV).*

Jesus came to enable man to be in covenant relationship with God by the grace of God through the avenue of faith. Jesus also came to show humanity what "*the Word becoming flesh*" looks like: It exemplifies Matthew 4:4, which says: "*Man shall not live by bread alone, but by every word that proceeds from the mouth of God.*" As Jesus spoke only words that the Father gave Him to speak, God's children may similarly occupy this state, because according to 1 John 4:17: "*As He is, so are we in this world.*" This state perhaps identifies our corporate full potential. We are to be the Word incarnate here on earth, having been patterned after Jesus—the Son of God, the Word of God.

Chapter 34
Absolute Value

That there are opposite forces in the world is no surprise to us. Whether negative-positive, south-north, up-down, cold-hot, or acid-alkaline, sustaining life as well as potential depends upon a careful balance, with the midpoint tending to serve as an ideal, such as the perfect alkaline-acid balance of 7. If we were to conceptualize these forces in terms of wave dynamics, the polar forces would climax at both the apex of a wave's crest and the point of a well's depression or lowest point on its trough. In this light, these points are simply two among many on a continuum with each having its own storage level of potential. However, other polar forces that require a slightly different illustration are the concepts of gravity and buoyancy, which, while opposites, are both the restoring forces of oscillations referred to as gravity waves.

Gravity waves take into account two types of waves—one that occurs on the surface of a medium, referred to as surface waves, and the other that occurs in the depths of the medium, referred to as internal waves. Both types of waves, however, exist between the forces of gravity and buoyancy, with both exerting pressure from different directions, and it is this median position that enables surface waves to be restored by either gravity or buoyancy. With the understanding that gravity pushes down to the earth and buoyancy keeps objects from drowning, what becomes apparent is that there is an inherent difference in the *intentionality* of each force when it comes to restoring the surface waves. For instance, if one were to consider surface waves as representing the movement of the Spirit of God, as it is illustrated in Genesis 1:2, *"And the Spirit of God moved upon the face of the waters"* (KJV), if the force of gravity were responsible for attempting to restore equilibrium, it would indicate a grieving of the Holy Spirit of God. If, however, buoyancy were the restoring force, due to the force of exertion coming from beneath and consequently "lifting up" the waves' troughs, it would signify that there was a release of the need to continue in prayer, praise, thanksgiving, or intercession. In essence, the force of buoyancy would indicate an attempt at restoration of peace.

With regard to surface water waves, for instance, these define the area where the water's surface and the atmosphere meet, dancing together as does positive space with negative space. Abiding in this space with connection as well as resistance, these waves may be most readily defined with respect to surface tension. While water molecules are equally pulled in all directions from their neighbors, the ones at the surface that don't have water molecules above them are pulled inward, which creates an inward pressure and forces the water molecules to contract. This forms a protective shield that keeps foreign molecules from easy entry, and as these molecules all adhere to one another, they corporately become easily swept up by incoming winds. Surface gravity waves then can be understood in terms of James 1:6-8: *"But let him ask in faith, nothing wavering. For he that wavereth [in faith] is like a wave of the sea driven with the wind and tossed. For let not that man think that he shall receive any thing of the Lord. A double minded man is unstable in all his ways"* (KJV). Words of the world, therefore, indicate the wind, and the waves that are driven by it and tossed are the individuals who have been corporately persuaded, as is readily seen when group think occurs.

In terms of surface tension, we see this play out in our social world all the time: People are drawn to connect with those who look and act just like themselves, and when they have *cohered* with one another, the tendency is to use this union as a shield to keep others from breaking through. The inherent judgment of this positional attitude keeps the closed-minded shielded from the truth and knowledge of love. However, if one were to stand in the deep spaces beneath the surface, they would stand under the surface tension, or in other words they might understand rather than judge and remain closed-minded.

Nonetheless, gravity waves are not strictly surface waves as mentioned; there are also internal gravity waves which occur within the depths of a fluid medium. Traveling at the interface between two fluids of different densities, such as oil and water, internal gravity waves are characterized by an intrinsic resistance, which may again be understood in terms of surface tension. With respect to internal water waves, which oscillate between waters of different densities, a phenomenon referred to as dead water may occur. When a thin layer of low density fresh water is above a thick, dense layer of salt water, a boat traveling therein may unexpectedly slow in response to this inherently resistant force. Directly related to James 1:8, the different densities relate to a double-minded man who travels as much in the flesh as in the Spirit, with the cares of this world as important as the purposes of the kingdom.

An additional Scripture applies, however, for the carnal-minded man who has also tasted the good gift of God (Holy Spirit) and travels between both the shallow and the deep. Hebrews 5:12-14 addresses such a state: "*For though by this time you ought to be teachers, you need someone to teach you again the first principles of the oracles of God; and you have come to need milk and not solid food. For everyone who partakes only of milk is unskilled in the word of righteousness, for he is a babe. But solid food belongs to those who are of full age, that is, those who by reason of use have their senses exercised to discern both good and evil.*"

Although not specifically addressing water in this context, Isaiah 42:5 strikes an interesting comparison between surface and depth: *"Thus saith God the Lord, he that created the heavens, and stretched them out; he that spread forth the earth, and that which cometh out of it; he that giveth breath unto the people* ***upon*** *it, and spirit to them that walk* ***therein****"* (KJV). God gives breath to all men, or to all children who walk *on* the surface of the earth. Those who remain on the surface, however, will be confronted with wind and waves, and like a wave of the sea they will be *"driven with the wind and tossed"* (James 1:6). However, those who exhibit depth not only are more buoyant—able to lift up descending troughs—but exist in a domain space that is hidden from the surface tempest.

This is where the restoring forces of gravity and/or buoyancy come in, and this actually provides an illustration for the exertion required to resist the anti-fruit of the Spirit, or less commonly to resist the fruit of the Spirit. The restorative forces of gravity and buoyancy can be likened to the concept of self-correction. For instance, if someone is primarily governed by the fruit of the Spirit (having a character that resists gravity and is consequently more buoyant) with truthfulness and honesty as fundamental guideposts, whenever the anti-fruit become dominant, a person so governed will tend to self-correct, where their actions and words will try to compensate for any offense. It is in this way that their self-image as a good and decent person may remain.

Adhering to the same pattern, however, if someone is governed by the anti-fruit and has as her self-image a tough woman who never gives in to other's demands, if a fruit of the Spirit leaks out and makes her feel weak, she too will self-correct and compensate by sporting arrogance and pride perhaps; this particular response, in turn, may be understood as being governed by the force of gravity. Anything that pulls one down and leads one toward operating according to the flesh would therefore be governed by gravity, and anything operating according to the Holy Spirit would be governed by the force of buoyancy. We can see by this explanation how it was that Jesus was able to walk on water: Father, Son/Word, and Holy Spirit are one and are governed by the force of buoyancy that opposes death and the grave.

As the force that corresponds with the anti-fruit of the Spirit due to its inability to promote freedom as well as its formidable resistance to it, the force of gravity keeps people grounded and earthly. Philippians 3:18-19 mentions the spiritually relevant gravitational pull: *"For many walk, of whom I have told you often, and now tell you even weeping, that they are the enemies of the cross of Christ: whose end is destruction, whose God is their belly, and whose glory is in their shame, who mind earthly things"* (KJV). Enemies of the cross of Christ, therefore, have their minds on the things of this earth, and it may be said that gravity has won, bringing into captivity every thought to the obedience of the antichrist, or to the local magnetic forces that operate on the earth. Anti-gravity or buoyancy, however, resists this gravitational force and pushes against the downward thrust with a force equivalent to the pressure from the depths below the surface and consequently enables "*bringing into captivity every thought to the obedience of Christ"* (2 Corinthians 10:5 KJV).

Interestingly, Jesus, the Word of God, has been referred to as the substance within which life exists, and that the life was the light of men: *"In Him was life, and the life was the light of men. And the light shines in the darkness, and the darkness did not comprehend it"* (John 1:4-5). All life forms, other than sea creatures and snakes, resist gravity. Life forms walk upon the earth in defiance of the downward thrust. Birds resist any gravitational pull in flight, and trees and plants break through often frozen ground to reach the sun. The sun (light) therefore, as a point attractor, helps life forms on earth overcome the gravitational force. The Son of God, Jesus, does so as well, providing those who believe with the potential of standing tall and walking with a force of equivalent value to the force of buoyancy.

If one were to consider gravity and buoyancy as two phases of a wave, gravity would be the trough, signifying a potential well, while buoyancy would be represented by the crest, the apex of which would signify the point of greatest potential energy. Understood in this context, the crest and the trough of a wave signify opposite forces and opposite domains. While connected at an equilibrium point, the domains never meet; the buoyant domain lifts up those who drift, while the gravitational domain pulls them down. And so are all those who do not make a choice to be continually lifted by the buoyant domain, the life that is by faith in Jesus Christ, the Messiah, who is our wisdom (see 1 Corinthians 1:24). "*If any of you lacks wisdom, let him ask of God, who gives to all liberally and without reproach, and it will be given to him. But let him ask in faith, with no doubting, for he who doubts is like a wave of the sea driven and tossed by the wind. For let not that man suppose that he will receive anything from the Lord; he is a double-minded man, unstable in all his ways"* (James 1:5-8).

That opposite forces exist in the world is not a question; however, somewhere along the road these opposite forces have come to be discerned through the lens of relativity. Similarly, and perhaps as a fundamental backlash to the notion of *absolute*—a value or principle regarded as universally valid or viewed without relation to other things—with regard to the truth of God, the theory of relativity has perhaps been applied inappropriately to the concept of absolute value. While *absolute* implies a principle that remains verifiable regardless of the measurement tools used, the definition of absolute value is a mathematical construction and is regarded "the value of a number *in relation* to its distance from zero."

Learning this concept early on in school, we were told to hone in on that zero point, a point on a number line, and we counted the numbers before and behind it. If we counted five points to the right of zero, we arrived at 5, and if we counted five points to the left of zero, we arrived at -5. Consequently, we learned that the five points counted were equivalent in value, where the absolute value of "5" as well as "-5" is the same: "5." Confined to the mathematical domain, this is a valuable means of measuring the values of distance. However, considering Einstein's theory of relativity, particularly special relativity, and how a point "zero" in any frame of reference is entirely arbitrary, as well as dependent upon the measuring instrument, this notion is suspect for unintentionally dulling the world's ability to discern true value.

While the applicable notion is obviously symmetry, there's one particular phenomenon that may apply more than other objects of symmetry—waves. The crest and the trough of a wave of water provide a perfect illustration of the concept of absolute value with the surface, imagined calm indicating point zero. The interesting thing about waves is that while we primarily consider only the crest of the wave, because the crest is the phase of the wave that we often see as it travels above the surface, there is a corresponding trough that travels beneath the surface. This phase mirrors the crest's amplitude under most conditions, except perhaps in shallow waters. However, it propagates in the opposite direction, toward the ocean floor instead of toward the sky. The fact is that the positive and negative amplitudes correspond, and these two phases can be understood in positive and negative terms, with this latter phase being considered negative. In light of this correspondence between the positive and negative amplitudes, the concept of absolute value makes not only mathematical sense but phenomenological sense as well.

Deductively then, if we were to draw a line through the plane that separated the phases at the exact point at which both the positive and negative amplitudes were of equivalent value, we would attain an illustration of absolute value, where the value between these two waveforms were equal even though one was positive and the other negative. Both wave *phases* would thus be identical in relative terms, but that's where their identical value would stop. While the crest would draw out a hill of sorts, representing a space upon which an imaginary object would have gained potential energy in its ascension, the trough would draw out a space that would epitomize a potential well, which is a space where an imaginary object would be stuck. In this position, without an influx of energy from the outside environment, the object would remain in its *stuck* state. The imaginary object would therefore be bound in a helpless position; it would be as one imprisoned. With consideration of this illustration, it seems fairly reasonable to assume that in truth the notion of absolute value is somewhat faulty.

Ascribing a relativistic value to something that can be easily illustrated according to natural law and which subsequently shows the breakdown of equivalent features seems to obliterate this concept's fundamental value. Nonetheless, what may be of equal value to these considerations is that while the knowledge of properties such as frequency and wavelength may be accepted as fact, I might suggest that due to their relationship to the fundamental operatives of the universe they be considered to be aspects of the *truth* instead. Whereas facts operate according to observables, truth operates according to principles that can be verified not only by natural occurrences, but whose premises can be extended beyond a one-dimensional platform. Principles, therefore, have an enduring quality. And with light being proven to operate as both a particle and a wave, it seems valid to assume that waves operate according to truth, with *light* light and *dark* light representing crest and trough respectively.

There appears to be a fundamental breakdown of soundness when an illustration of a concept backs the essence of the concept, but clearly demonstrates that the principle that it stands for has no validity. I can only imagine that the origin was related to Newtonian physics, which was based on reversible processes that are cyclic in nature. In cyclical processes, no great amount of energy is lost, having naturally rebounded from any enduring decay. Fulfilling the notion of God as unchanging, for centuries Christians have gravitated to the notion of reversibility as epitomizing God. However, in actuality, God's ways more accurately reflect the systemic transformational processes that occur after the fulfillment of a system's decay. Ilya Prigogine addresses this notion in his theoretical outline of dissipative structures, where evolutionary or creative energy is brought forth out of that which has decayed.

It is evident that both systems are present in the universe; however this system, which corresponds with entropy and our experiential understanding of time as a one-way arrow, replicates God's fundamental way of fulfilling His Word. God is the Master of renewal, of rebirth, of bringing dead bones to life. While in an ideal world the idea of reversibility is grand, it's not the basis of the greater part of reality. Outside of the systems which operate, for example, according to the principles of motion that apply to large bodies governed by gravity, the majority of systems operate in alignment with irreversibility and one-way entropy. However, in light of Prigogine's dissipative structures, it might only require time before the cyclic nature of life actually manifests; as we are in this solid phase for only a finite amount of time, we may not have the ability to assess the cycle accurately.

With regard to the experiential application of the concept of absolute value, psychological processes come to mind. Based on the notion of positive and negative amplitudes—hills and wells, so to speak—we see the real life effect of these positional phases. Many people are in a positional space equivalent to a potential well. They are stuck in their attitudes, interpretations, and thought processes; they require an outside force that will provide them with the momentum to release them from their bound state. In other words, in order to overcome the presenting barriers, coincidentally related to *fact* as opposed to the *truth*, only an outside force is capable of ushering them out of their *stuck* state and up into a realm of freedom. Jesus, the Word of God, is such a force; His Word provides the outside force that enables ultimate freedom.

Simply stated in Psalms 146:7: "*The Lord gives freedom to the prisoners.*" Galatians 5:1 pronounces: "*Stand fast therefore in the liberty by which Christ has made us free,*" And in Isaiah 58:6, a question regarding His overall purpose is posed: *"Is this not the fast that I have chosen: to loose the bonds of wickedness, to undo the heavy burdens, to let the oppressed go free, and that you break every yoke?"* If this was indeed His apparent commitment, which has been supported not only in Scripture but also in the personal testimonials provided by millions whose lives have been released from poverty, sickness, lack, depression, and hopelessness, is it within reason to doubt that He was and is capable of fulfilling His commitment? As we are well aware in the natural realm, when someone is not faithful to his word and does not perform that which he has spoken, how long does it take us to abandon our hopes and faith in that person? Not long at all. And yet, it's more than two thousand years since Jesus died for us, and still there are millions who trust Him. I might suggest that they know something that perhaps others don't.

Chapter 35
The Kingdom of God System

"See, saith [God], that thou make all things according to the pattern shewed to thee in the mount" (Hebrews 8:5 KJV). Intricately tied to our lives are patterns of existence that have been observed and consequently have enabled us to study them and obtain knowledge of ourselves as well as our world. In economics, psychology, and politics, the patterns within the systems under study are sifted through as with a fine toothed comb, until man feels that he has obtained some level of control over the systems under study; knowledge is power, and as a species that feels basically powerless from a macro-perspective, man needs to feel dominant and in control. This too though is just a pattern discerned from observation. However, one thing remains beyond man's control, for "*The kingdom of God cometh not with observation: neither shall they say, Lo here! or, lo there! for, behold, the kingdom of God is within you*" (Luke 17:20-21 KJV).

Things pertaining to God cannot, therefore, be measured with any man-made measuring tool; neither microscope nor telescope, neither sonogram or x-ray machine, nor even the monstrous Large Hadron Collider can enable man to gather tangible evidence of the invisible God. However, similar to the way in which the positron, the electron's anti-particle, was hypothesized to exist through the patterns of motion established by Isaac Newton, the forces that power God can be discerned through their effect on matter. The patterns of the visible, or that which is known to exist, then are able to point the way toward the invisible, as Romans 1:20 attests: *"For since the creation of the world His invisible attributes are clearly seen, being understood by the things that are made."* All of the systems, processes, and patterns that man is able to discern through observing the physical substance of the world, therefore, have first been *created* in the conceptual realm and are then utilized by both man and God to *make*, or bring to fruition, substance in the physical realm.

Among the established patterns to which the world adheres is the tendency for all systems to seek equilibrium and simultaneously respond to the influxes of change in environmental conditions; this, of course, parallels the widely-accepted theory of evolution. And all observations and data acquired to support that theory are by all accounts accurate, and most importantly are relevant to the phase transitional patterns that were preordained, but not by nature or by the universe, but by God Himself, the Father of Abraham, Isaac, and Jacob. "*God, who at sundry times and in divers manners spake in time past unto the fathers by the prophets, hath in these last days spoken unto us by his Son, whom he hath appointed heir of all things, by whom also he made the worlds*" (Hebrews 1:1 KJV). "*In the beginning was the Word...and the Word was God. All things were made through Him, and without Him nothing was made that was made*" (John 1:1,3).

Of vital importance is the understanding that everything in this earth is truly created in the image of already-established substance in the heavenly realm. Plato (427-347 B.C.) was the first to apprehend this truth in his theory of ideal forms, and many others have touched upon this notion. Why? Because it's the truth, as is the power of visualization: "*Faith is the substance of things hope for, the evidence of things not see.*" Faith requires visualizing the end result of what one hopes for, as is confirmed by Proverbs 29:18: "*Where there is no vision, the people perish*" (KJV). Accordingly, understanding what God has said provides a picture of the desired end, which is exactly what God had done when He created the worlds: His creation of heaven and earth and all that was therein existed in the potential form of a vision or conceptualization.

The substance of what God had desired to come to pass, spoken about in Genesis 1, was not manifested in the physical realm until He had caused a mist to come up from the earth. Even so, in that first chapter after His creations were fully conceived each day, "*He saw that it was good*" or He saw that His creation would serve the desired purpose or end. He had faith, or "*the substance of things hoped for, the evidence of things not seen*" (Hebrews 11:1) that His visions as well as purposes would materialize and be fulfilled. The process of conceptualization thus functions as a phase, which transitions into a manifesting phase as a result of faith having been consistently applied.

However, applying the notion of a phase transitional stage between an image and the actual substance is beyond our natural inclination, because when we get an idea in our minds, if we want to see it actualized we have been indoctrinated into the belief system that we must *work* for it or *walk it* out. And this is where it becomes difficult, because while we must work, pushing against resistant forces that keep us bound and keep us from fulfilling our God-given potential, God's perfect will for us is to have us enter into His rest (see Hebrews 4:3). This entire thing becomes further complicated when local magnetic forces remind us of a familiar saying: "God helps those who help themselves." This, by the way, does not adhere to God's Word; it does, however, help to build up strongholds in our lives. And as a result, we remain bound to the idea that anything that is possible *is* possible only because of personal effort or luck.

The mindset that takes hold as a consequence of these strongholds may be understood as reflecting the degree to which we are limited by a first order phase transitional understanding of evolutionary processes. Bound by natural understanding, it reflects the need to assess all intermediate phases and consequently rejects a fundamental belief in the extending power of second order phase transitions, which occur in an instant and cannot be readily measured. In essence, the difference may be assessed according to the difference between flesh and Spirit, which ultimately relates to death versus life according to Romans 8:6: *"For to be carnally minded is death, but to be spiritually minded is life and peace."* As a result of this thought process, the anti-fruit, which are inherently contrary to the fruit of the Spirit, enter in and bind our minds' ability to comprehend the delimitations of the Spirit. Envy, drunkenness, strife, hatred, pride, impatience, and the like are born out of this bound and limited understanding of the realm of possibility.

Noah and Abraham and Moses, however, are three examples of men who didn't *walk it out* by themselves; rather they walked with God and listened to His commands and prospered, *"For with God nothing will be impossible*" (Luke 1:37). According to Webster's dictionary, the verb "to walk" means to "move along on foot or to advance by steps[7]," and this is exactly what these three men did; they advanced in life by taking steps, one at a time, until they had accomplished God's mission and vision for their lives. While Moses, who died in faith, never made it to the Promised Land due to a moment of rashness and disobedience (see Numbers 20:11-12), Noah and Abraham received the promises given to them by God because of their faith, patience, and obedience to His commands. These three men, as well as all who are mentioned in Hebrews chapter 11, have consequently become for us a cloud of witnesses to the truth as Hebrews 12:1 proclaims, and have become examples to follow as Hebrews 6:12 commands, *"imitate those who through faith and patience inherit the promises."*

As these faithful followers of God serve as examples of witnesses to God's truth on earth, an inherent resonation with the other witnesses—the Spirit, the water, and the blood (see 1 John 5:8)—on earth is implied. And these, in turn, resonate with the record-bearers in heaven, which according to 1 John 5:7 are "*the Father, the Word, and the Holy Spirit, and these three are one."* Anyone who has receptor seeds for the Word of God in their heart will consequently resonate with both the record-bearers in heaven as well as the witnesses in earth, and such resonation will bear witness to the truth.

As the processes and patterns apparent in this natural realm appear to correspond with heavenly patterns, we may infer that there is reason and purpose for this correspondence, despite the cultural lure of meaninglessness. While

7 *Webster's Collegiate Dictionary*, s.v., "walk."

some may maintain that existence is fundamentally meaningless and without purpose, the bridges between science and the Word of God appear to contradict this assumption. And if the integrity of these bridges remains and is not shaken like those things that are made (see Hebrews 12:27), then one purpose indeed exists, and that one purpose is to fulfill God's Word. For what particular reason? So that His kingdom will come and His will shall be done on earth as it has been patterned in heaven (see Matthew 6:10). Accordingly, one of God's desired ends was to provide all men, not just Israel, His chosen people, a way to be redeemed. Israel had been given the Law, which had become a principle upon which God had bound Himself. One important aspect of the Law, which was the glue of the covenant between God and man, was the shedding of blood of particularly prized firstborn animals without spot or blemish. Therefore, in order to maintain His integrity, it became necessary for Him to instate a way by which all men might be able to *pay* for their sins; this, of course, required shed blood, which is why circumcision became a sign of Israel's covenant with God.

Hebrews 9:22 mentions the importance of blood being shed for the remission of sins: "*And according to the law almost all things are purified with blood, and without shedding of blood there is no remission.*" And, "*So Christ was offered once to bear the sins of many. To those who eagerly wait for Him He will appear a second time, apart from sin, for salvation*" (Hebrews 9:28).

> *For if the blood of bulls and goats and the ashes of a heifer, sprinkling the unclean, sanctifies for the purifying of the flesh, how much more shall the blood of Christ, who through the eternal Spirit offered Himself without spot to God, cleanse your conscience from dead works to serve the living God? And for this reason He is the Mediator of the new covenant, by means of death, for the redemption of the transgressions under the first covenant, that those who are called may receive the promise of the eternal inheritance (Hebrews 9:13-15).*

This, therefore, reflects a grand purpose of God, who is forever merciful to the children of men. Nonetheless, utilizing the principles of phase transitions, where a physical substance can transform into another, as well as atomic decay, where elements assume a new identity, God has provided a way to fulfill His will without operating outside His principles or laws. This is key, because God cannot operate outside of His already established covenants with the heavens and the earth, because that would make him a liar, and "*God is not a man, that He should lie, nor a son of man, that He should repent. Has He said, and will He not do? Or has He spoken, and will He not make it good?*" (Numbers 23:19). In other words, God has a purpose, a vision of the desired end, which is aligned with His will, which is revealed in His Word. And while the intermediate phase state between image and creation remains elusive, it is not beyond the human grasp.

Ilya Prigogine and Isabelle Stengers in their work, *Order Out of Chaos,* allude to a phase transitional state that occurs after systems have undergone entropic decay. After a period of delay, the decayed system actually rebounds and begins to develop new features, whereby it may be said that the structure of the system has *evolved.* Although understood in evolutionist terms, the dissipative structures spoken about present an accurate portrait of the pattern of phase transitions that were preordained by God. As an example, the final book of the Bible reveals what will occur at the very end of this dispensation:

> *And God will wipe away every tear from their eyes; there shall be no more death, nor sorrow, nor crying. There shall be no more pain, for the former things have passed away." Then He who sat on the throne said, "Behold, I make all things new." And He said to me, "Write, for these words are true and faithful." And He said to me, "It is done! I am the Alpha and the Omega, the Beginning and the End. I will give of the fountain of the water of life freely to him who thirsts (Revelation 21:4-6).*

From this Scripture we may ascertain that Prigogine and Stengers' dissipative structures do indeed emulate God's system of evolutionary processes. And we may discern a transitional process as having occurred between the time that Adam walked on earth and the time when Jesus will return.

> *And so it is written, "The first man Adam became a living being." The last Adam became a life-giving spirit (1 Corinthians 15:45).*

> *The first man was of the earth, made of dust; the second Man is the Lord from heaven. As was the man of dust, so also are those who are made of dust; and as is the heavenly Man, so also are those who are heavenly (1 Corinthians 15:47-48).*

> *Flesh and blood cannot inherit the kingdom of God; nor does corruption inherit incorruption. Behold, I tell you a mystery: We shall not all sleep, but we shall all be changed—in a moment, in the twinkling of an eye, at the last trumpet. For the trumpet will sound, and the dead will be raised incorruptible, and we shall be changed (1 Corinthians 15:50-52).*

> *So when this corruptible has put on incorruption, and this mortal has put on immortality, then shall be brought to pass the saying that is written: "Death is swallowed up in victory" (1 Corinthians 15:54).*

These Scriptures thus allude to the ultimate phase transition—the phase transition from an earthly body to a heavenly body, through the patterned process of sublimation.

Chapter 36
Internal Gravity Waves

As the patterns and processes visible on earth have their origin in heavenly places, there is an intrinsic connection that may be understood according to the phenomenon of internal gravity waves. Genesis 1:6-8 alludes to this feature: "*Then God said, 'Let there be a firmament in the midst of the waters, and let it divide the waters from the waters.' Thus God made the firmament, and divided the waters which were under the firmament from the waters which were above the firmament; and it was so. And God called the firmament Heaven.*"

From this Scripture, we can discern that internal gravity waves actually define this firmament, which is, in turn, referred to as heaven. Of course, let us not forget that this first chapter of Genesis deals with the *creation* of the heavens (plural) and the earth, and not the formation, which occurred at a later time. Nonetheless, as internal gravity waves separate two fluids of different densities, the firmament mentioned appears to do precisely that; it divides the waters which were under the firmament from the waters above the firmament (see Genesis 1:7). And therefore this realm may be said to be simultaneously influenced by both realms' movement and yet remains incapable of penetrating either. Luke 16:26 mentions this firmament: "*And beside all this, between us and you there is a great gulf fixed: so that they which would pass from hence to you cannot; neither can they pass to us, that would come from thence*" (KJV).

In reference to a conversation that Abraham had with a rich man who had had no sympathy for a beggar named Lazarus and had gone to hell, while he was conscious of the divide as well as the reality of the domain in which Abraham resided, the firmament between them was impermeable and consequently resisted both domains. This particular firmament would therefore signify a domain space between God's kingdom of heaven and the lake of fire, or the center of the earth that is filled with magma. Confirmed by the rich man's mention of being "*tormented in this flame*" (Luke 16:24 KJV), Revelation speaks of this domain: "*These both were cast alive into a lake of fire burning with brimstone*" (Revelation 19:20 KJV).

This particular realm that parallels internal gravity waves, as well as the firmament that separated the waters, rides fluidly upon the denseness of hell and beneath the lighter waters of heaven. And while the kingdom of heaven appears to refer to a system that counters the systems of this earthly realm, the firmament, referred to as "heaven" in Genesis 1:8, seems to refer to the atmosphere above the earth. And it is within *this* realm where all of the frequencies available to man reside. In this domain, both the frequencies that are generated by the local magnetic forces that persuade men to reject the one true global magnet—God—as well as the frequencies which correspond with the fruit of the Spirit sojourn. Duly accessible to human beings who are electrically charged, having electrolytes coursing through their veins, these frequencies readily attach themselves to those who are predisposed; those who have within their hearts receptor seeds for locally *generated* frequencies will receive of these *spirits*, and those who have receptor seeds for the Word of God and His Spirit will receive of these instead.

In essence this *heavenly* domain, which parallels internal gravity waves, is a field replete with opposing as well as complementary charges. Depending upon individual charges, which are determined by the receptor seeds within the heart, both attractions to as well as repulsions from specific frequencies that have been captured by men are experienced. Herein Ephesians 6:12 becomes more plainly understood: "*For we do not wrestle against flesh and blood, but against principalities, against powers, against the rulers of the darkness of this age, against spiritual hosts of wickedness in the heavenly places.*" Men are not our enemies; however, the frequencies/spirits to which they have connected are.

When accepted into people's lives, these particular frequencies effectively refract (send off in another direction) God's white light frequency of love, which exists beyond this particular realm. As electrical fields are perpendicular to magnetic fields, this frequency of love travels perpendicular to the earth's surface, which serves as the source of the local magnetic field; with the earth's plane constituting the magnetic field, the white light frequency travels at a ninety-degree angle of pure "*righteousness*." As righteousness comes by faith (see Romans 3:22), it is consequently through this one frequency whereby the grace of God may be imparted to men as Romans 5:21 alludes: "*That as sin hath reigned unto death, even so might grace reign through righteousness unto eternal life by Jesus Christ [the Word] our Lord*" (KJV). As grace reigns through righteousness, therefore, grace is coupled with this frequency of love. And when any light travels through two fluid bodies of different densities, such as internal gravity waves, there is a consequent refraction of the light, with the density of the bottom fluid determining the degree to which the light bends.

Accordingly, righteousness signifies a condition where when the white light of God's love passes from one medium to another, or from God to man or from one person to another; instead of being refracted because of the density of the receiver, it passes without refraction. This lack of deflection, therefore, signifies that the one to whom the light passes has no significant density, or may be said to respond to the light with pure reflection, or rather love. As righteousness comes by faith, the white light frequency of God's love must be received by faith. The avenue of faith—which has been made available by God's gift of grace—is the conduit whereby men are able to receive the love of God and thereby experience salvation: "*For by grace you have been saved through faith, and that not of yourselves; it is the gift of God*" (Ephesians 2:8). Righteousness, therefore, simply signifies one's ability to absorb and reflect back the white light of God's love, which is enabled by grace through faith.

Chapter 37
Grace and Glory

By God's grace and His grace alone do all people have access to the ninety-degree righteous white light of God's love through the avenue of faith in Jesus Christ, the Word of God. By God's grace, therefore, all men, Jews and Gentiles, have been given access to the promises of God that had been earmarked for Israel alone. Due to their fathers' disobedience of not fully trusting God, in addition to becoming mixed with other men's gods, the Gentiles were given an opportunity, through the avenue of faith, to become God's people, as Romans 11:11 suggests: "*through their fall salvation is come unto the Gentiles, for to provoke them to jealousy*" (KJV). So, if it was grace that enabled *all* men to be saved because of the chosen ones' disobedience, then God's grace simply performed the function of extending the call to obedience to others besides the Jews. Describing this one vital way of God's kingdom of heaven, the parable of the marriage of the king's son in Mathew 22:2-3 delineates for us a pattern of the kingdom of heaven's system: "*The kingdom of heaven is like unto a certain king, which made a marriage for his son, and sent forth his servants to call them that were bidden to the wedding: and they would not come*" (KJV).

As the king's servants went forth, so too does the frequency of God's white light, searching for someone who has a receptor seed for the light in his/her heart. Those in the parable who would not come when called had no receptor seed for the king's words, and rather were drawn to the local magnetic forces, such as the callings from their work or from their needs and desires of goods and services. The parable, however, continues with the king sending soldiers to kill the offenders.

With the realization that the original chosen ones had no interest in attending an event that the king himself cherished, the king went into every unsuspecting corner of the country and called those who might be more responsive and appreciative of His offer. And lo and behold, when he had made his offer available to others, they gladly accepted. In this parable, the original wedding guests represent the Jewish people who were first called to accept Jesus. When they rejected Him the call went out to the Gentiles. And the Gentiles represent all those who were called just because the Jews didn't accept God's offer. This is the grace that God has given those who may accept the call to obedience to follow after righteousness.

Consequently, while grace has been communicated by some to exist in a polar opposite domain from the law, with the law being portrayed as a simple confinement or bondage while grace sets one free, the two are actually positioned perpendicular to one another with the intention of fulfilling righteousness. According to Matthew 5:17-18, Jesus came not "*to destroy the law, or the prophets: I am not come to destroy, but to fulfil. For verily I say unto you, Till heaven and earth pass, one jot or one tittle shall in no wise pass from the law, till all be fulfilled*" (KJV).

As the Word of God, Jesus—who became the Mediator of the New Testament and consequently of grace, through faith in Him—had as His fundamental purpose to speak words that would bear witness of the Law and the prophets, and thereby fulfill them. Grace, therefore, is a gift of God by which all men have been enabled access to the righteous white light of God's extending love. However, what is required to be granted this access is the acceptance of the invitation, which requires faith, "*for he who comes to God must believe that He is, and that He is a rewarder of those who diligently seek Him*" (Hebrews 11:6).

Existing primarily for the purpose of extending the promises of God to others besides God's originally chosen people, grace in and of itself is not an operational mode which can be accessed by our own conscious effort, as is faith. While we access faith "*by hearing, and hearing by the Word of God*" (Romans 10:17), grace is a gift that remains in God's utter control; it is reserved for those who have been called to obedience and have in turn obeyed the call. Those who are called and respond with obedience may be regarded as servants as well as sons who don't think more of themselves than they do their Master, and for them grace is reserved, as 1 Peter 5:5 declares: "*God resists the proud, but gives grace to the humble.*"

First Peter 5:10 addresses the true grace of God wherein we stand: "*But may the God of all grace, who called us to His eternal glory by Christ Jesus, after you have suffered a while, perfect, establish, strengthen, and settle you.*" From this Scripture, we see that grace enables us to endure, and if we endure, then we will overcome this world, as 1 John 5:4 says: "*And this is the victory that has overcome the world—our faith.*" And according to Romans 5:1-5: "*Therefore, having been justified by faith, we have peace with God through our Lord Jesus Christ, through whom also we have access by faith into this grace in which we stand, and rejoice in hope of the glory of God. And not only that, but we also glory in tribulations, knowing that tribulation produces perseverance; and perseverance, character; and character, hope. Now hope does not disappoint, because the love of God has been poured out in our hearts by the Holy Spirit who was given to us.*"

The grace, therefore, wherein we stand allows us to endure with faith, patience, and hope for the glory of God. This, consequently, makes the glory of God that which 1 Corinthians 9:24-25 refers to as the prize for which men compete: *"Do you not know that those who run in a race all run, but one receives the prize? Run in such a way that you may obtain it. And everyone who competes for the prize is temperate in all things. Now they do it to obtain a perishable crown, but we for an imperishable crown."* The incorruptible crown that we run for is thus the glory of God, which appears to correspond to second order phase transitions.

As a heavenly pattern that demands earthly reflection, the glory of God is manifested on earth in natural phenomena. Similar to the instantaneous effect that a permanent magnet has on the atoms in a paramagnet, setting them in order in a twinkling of an eye, glory also appears instantaneously. One such natural manifestation is a rainbow, which besides being patterned after God's glory is also a means by which God's Word is glorified. Occurring in the visible realm, the spectrum of light that occurs in a rainbow is caused by the sun's rays shining behind droplets of moisture at a low altitude angle in the atmosphere.

While the occurrence of a rainbow is understood as both a meteorological and optical phenomenon, its appearance became coupled with a covenant that God made with the earth for Noah's sake. Because of the fact that "*Noah found grace in the eyes of the Lord*" (Genesis 6:8), which was provided him as a gift, due in part to his obedient heart, a strong bond of a covenant (like the mussel and the rock) was established, with the rainbow as a sign: *"'The rainbow shall be in the cloud, and I will look on it to remember the everlasting covenant between God and every living creature of all flesh that is on the earth.' And God said to Noah, 'This is the sign of the covenant which I have established between Me and all flesh that is on the earth'"* (Genesis 9:16-17).

While the rainbow is a visible *token* of God's grace, a bright halo is often associated with an expression of God's presence or His glory, as it signifies the deposition of the Spirit of God. Special manifestations of light phenomena appear as essential patterns that God has used to exhibit His presence in and on the inhabitants of the earth with an apparent purpose that they will know that He is Lord. As an example, a visible manifestation of God's emanating light whereby He may be glorified is referred to in Matthew 5:16: "*Let your light so shine before men, that they may see your good works and glorify your Father in heaven.*" Essentially, when we connect with God's white light frequency of love that intercepts the earth's plane at a ninety-degree *righteous* angle, we absorb and reemit that light as opposed to absorbing or refracting it. With this light emanating out of our hearts, we are thus accounted as He is in this world (see 1 John 4:17).

However, as we are accounted righteous through faith by grace, the faith that essentially enables reception of the white light of God's love is nonetheless demonstrated by accompanying action. And this action is paramount in glorifying God, who bestows righteousness upon man. James 2:21-23 confirms this: *"Was not Abraham our father justified by works when he offered Isaac his son on the altar? Do you see that faith was working together with his works, and by works faith was made perfect? And the Scripture was fulfilled which says, 'Abraham believed God, and it was accounted to him for righteousness.' And he was called the friend of God."* Abraham proved that he believed God by taking action when there was no proof that it would turn out in Abraham's favor. In the natural realm, his actions looked insane, but Abraham obeyed God anyway, and his actions signaled to God that he trusted Him and could in turn be trusted.

While works or actions can certainly signal faithfulness, alone they fail to secure anything other than prideful confidence. Likewise, faith alone falls short of transitioning into the *"substance of things hoped for, the evidence of things not seen"* (Hebrews 11:1). Why? Works allow for the proving of our faith in God, just as David, having faith in God's ability to deliver his enemy into his hand, first proclaimed to Goliath that he would kill him; in other words, he called those things that were not as though they were (see Romans 4:17), demonstrating faith, and then he armed himself with his sling, a weapon he had used before to kill a lion and a bear, demonstrating action. Having faith in God's power, ability, and intention to back His covenants, David, like Noah and Abraham, had proven his faith in God's grace, which enabled them all to operate by faith, by first stepping out in the natural realm and taking action. It is this process that ensures that God is glorified, which, after all, is what it's all about.

Faith is therefore the governing force behind the works alluded to; therefore the manifestation of light that emanates from those who might glorify God comes from the abiding white light of love that opposes the darkness or fear of this world. Matthew 5:16 alludes to a visible sign of God's presence that according to Matthew 6:21-22 is indicative of the treasure that one has within his/her heart: *"For where your treasure is, there will your heart be also. The light of the body is the eye: if therefore thine eye be single, thy whole body shall be full of light"* (KJV). Of potential significance, "single" relates to single-mindedness as well as to "folded," which implies a certain vigilance or protectiveness about what one lets into one's heart. With potential reference to protein folding, a process through which the polypeptides fold into their three-dimensional functional structures from their unfolded state or random coiling, singleness, or foldedness, implies that fulfilling one's potential as an emitter of light is contingent upon adhering to the preordained sequence and order for one's life.

While the glory of God appears instantaneously and is something for which we rejoice in hope as every heavenly pattern is made manifest here on earth, what earthly phenomenon parallels the glory of God? That glory is signified by a light that appears to display a greater vibrational character suggests that the pattern of the process involved would require light absorption at a different and/or higher level than normal. While under most circumstances, when a photon is absorbed by an electron within a non-metallic, non-conductive atom, the energy states of both are very similar; precisely, the photon's energy matches the difference between the electron's energy state and the adjacent energy level. In other words, the electron *resonates* with the photon's energy level and its particular frequency, and the atom emits a photon of this same energy difference. Another process of light emission, however, fluorescence, occurs when a higher energy photon excites an electron of a lower energy state; the electron is excitedly stimulated by a frequency of light greater than its own, and that frequency of light is completely absorbed.

As with a high energy child who brings a quieter one out of his shell, the bound electron (thought) will jump to a higher energy state in response. Instead of just dropping down to an adjacent lower level when the excess energy (photon) is emitted in the eventual process of seeking an equilibrium state, the electron relaxes all the way to its ground state when it releases its light. In other words, the reception and absorption of the light induces rest as well as enables the reemitting of the absorbed light at a slower and more enduring pace. Emulating "*the peace of God, which surpasses all understanding, will guard your hearts and minds through Christ Jesus*" (Philippians 4:7), this fluorescent state epitomizes the process of entering into the rest of God (see Hebrews 4:11). This enduring ground state of rest, not surprisingly, is found in the closest electronic energy level to the nucleus, which is the atom's life force. Interestingly, when an atom undergoes the process of beta-plus decay, whereby it assumes a new identity, it is from this inner shell that an electron is captured. Whatever is close to our heart, therefore, will determine the light that constitutes our inner life force.

Of equivalent value and function to this life force is the heart of man, which—as a conductor of the electrolytic fluid of blood, in which is life (see Leviticus 17:11)—is the conductor of the electrical signals that correspond with a light frequency and determine a man's life. This, in turn, provides an illustration of how important our *connections* are; they determine whether our paths will lead toward life or whether they will lead toward death, as Deuteronomy 30:19 proclaims: "*I call heaven and earth to record this day against you, that I have set before you life and death, blessing and cursing: therefore choose life, that both thou and thy seed may live*" (KJV).

As the heavenly pattern of fluorescence appears to be the process by which a man may obtain life and peace, and by which the Spirit of glory may rest upon him (see 1 Peter 4:14), the difference between that which is absorbed by an electron (thought) and its original energy level becomes equivalent to the potential difference or *voltage* of glory. This is where glory and grace have their greatest connection: While the spiritually mature will walk with God and experience His glory because of the close communion, the smaller the difference between a man's level of *spiritual enlightenment* (energy level) and God's level of being (energy level), the smaller the *voltage* of glory. However, the less *spiritual* a man may appear to be, when the Lord touches him with His light, His grace (unmerited favor) will enable a greater *voltage* of glory. As the grace of God is a gift in God's full control, He gives it freely to whomever He wishes. And here we have an explanation of why the Lord may have said, *"But many that are first shall be last; and the last first"* (Mark 10:31 KJV). Those who are more mature experience less wiggle room, for "*unto whomsoever much is given, of him shall be much required*" (Luke 12:48 KJV).

As God's domain is the entire electromagnetic spectrum, this process corresponds perfectly with what appears to occur as God's higher energy frequency interacts with our limited ones. Immediately overpowered by the absorption of His higher frequency sources of light as they are received or deposited upon us, we are brought to a ground state corresponding to what occurs when one *falls* under the power of God's Holy Spirit. As a response to this higher frequency of light being absorbed by electron(s) of a slower frequency, the reemitted light surpasses normal light's vibrancy, but at the same time is of a slower, longer wavelength, which corresponds with the peace of God. And herein lies a description of the glory of God, after which the phenomenon of fluorescence has been patterned; it alludes to God's supernatural light being emitted from within our limited frames and providing us with an impartation of His nature. In that the emitted light is of a slower, longer wavelength, it corresponds with the peace of God: *"But may the God of all grace, who called us to His eternal glory by Christ Jesus, after you have suffered a while, perfect, establish, strengthen, and settle you*" (1 Peter 5:10). "*And the peace of God, which surpasses all understanding, will guard your hearts and minds through Christ Jesus*" (Philippians 4:7).

As the path that is traveled from the invisible to the visible can be verified by man only with the aid of high-powered instrumentation, the mediating domain must otherwise be considered along relative lines. While man is limited through the fleshly sense level and is able to understand and know only that which has been given him, God sees all. While our limitations are spoken about in 1 Corinthians 13:12, "*For now we see in a mirror, dimly, but then face to face. Now I know in part, but then I shall know just as I also am known,"* God's unlimited perspective appears in Hebrews 4:13, "*all things are naked and opened unto the eyes of him with whom we have to do*" (KJV).

As Jesus is the Word and is One with God, the power of the Word is as an x-ray: "*For the word of God is living and powerful, and sharper than any two-edged sword, piercing even to the division of soul and spirit, and of joints and marrow, and is a discerner of the thoughts and intents of the heart*" (Hebrews 4:12). Nonetheless, that which is invisible to the human eye very often does display its existence by having an effect on the visible. "*For since the creation of the world His invisible attributes are clearly seen, being understood by the things that are made, even His eternal power and Godhead*" (Romans 1:20). While we are not able to see the incident ultraviolet radiation that is absorbed by an object that fluoresces, we are able to see the effect of its absorption, i.e., the reemitted light, "*in the twinkling of an eye*" (1 Corinthians 15:52); the light assumes a new nature.

These processes signify what occurs when the light of God's Word strikes a person at a heart level, causing the Spirit of God to "fluoresce," so to speak. While fluorescence signifies the glory that appears when we are in the presence of the Holy Spirit of God, when we are *knocked out* or *under the power* of a particular anointing, phosphorescence, which led to the discovery of radioactivity in 1896, signifies a delayed and yet more enduring glory.

Unlike fluorescence, a phosphorescent object reemits the absorbed electromagnetic radiation very slowly and after a delay—a period that signifies that *work* needed to be accomplished in the invisible realm. Reemitted at a lower intensity than the original excitation, the pattern of an object exhibiting phosphorescence, like fluorescence, signifies that absorption of the light of God has caused peace to rest upon the object. However, with the ground state or equilibrium state as a point attractor, a point to which systems tend, which elicits order out of chaos, phosphorescence is indicative of a more enduring state of glory, sometimes lasting for hours. Implicit in both fluorescence and phosphorescence, however, is an excited vibrational mode that occurs in response to a light of greater frequency, which in turn causes a remittance of light with more intensity. In sum, the greater the absorption of the light, which corresponds with having more electrons (thoughts) absorbing more Word, the greater the ability to return to a ground state of rest while at the same time evidencing an effectual glow.

Numbers 6:22-26 supports the connection between grace and glory as a gift from God: "*And the Lord spoke to Moses, saying: 'Speak to Aaron and his sons, saying, "This is the way you shall bless the children of Israel. Say to them: 'The Lord bless you and keep you; the Lord make His face shine upon you, and be gracious to you; the Lord lift up His countenance upon you, and give you peace.'"'*" God's light emission, being that much greater than ours, causes us to luminesce when we absorb His frequencies; by reemission we are able to exhibit the glory of God similar to what occurred to Moses when he descended Mount Sinai: "*Now it was so, when Moses came down from Mount Sinai (and the two tablets of the Testimony were in Moses' hand when he came down from the mountain), that Moses did not know that the skin of his face shone while he talked with Him. So when Aaron and all the children of Israel saw Moses, behold, the skin of his face shone, and they were afraid to come near him*" (Exodus 34:29-30).

This was an example of an impartation of God's glory, which was related to God's light being deposited in Moses in a special way, and the contingency factor for this impartation is notably the relationship held between God and, in this case, Moses. Exodus 33:11 provides an illustration: "*So the Lord spoke to Moses face to face, as a man speaks to his friend,*" which implies that Moses *walked* with God by faith, just as Noah and Abraham had—in close fellowship. So, the manifestation of the glory of God appears to be coupled with the level of intimacy that one has with God.

Distinguishable only by the delay in reaction, both fluorescence and phosphorescence are luminescent reactions to external light, which serve only as patterns that exemplify the glory that is spoken of in Revelation 21:23: "*The city had no need of the sun or of the moon to shine in it, for the glory of God illuminated it. The Lamb is its light.*" The Lamb is Jesus, the Word of God, who is the light of men. The Word of God, therefore, may be understood to be the source of luminescent power, which is the glory of God.

As the end result of an object that luminesces is the object subsiding to a restful state, so too is the end of the man who has been touched by the glory of God. Exodus 33:13-14 provides an example of when Moses spoke to the Lord and reminded Him of what He had said: "*'Yet You have said, "I know you by name, and you have also found grace in My sight." Now therefore, I pray, if I have found grace in Your sight, show me now Your way, that I may know You and that I may find grace in Your sight. And consider that this nation is Your people.' And He said, 'My Presence will go with you, and I will give you rest.'*" God's presence, His light, thus provides both peace and rest for those who are in resonant communion with Him.

That rest coincides with God's plans and purposes is not only evidenced by conservation laws, where systems are inherently restored to an equilibrium state, but also through processes whereby matter as well as light transcend their present state. As the glory is signified by fluorescence and phosphorescence, the message here is that God wants us to respond to Him, to His Word, and to His Spirit with an equivalent resonation. That the human race resonates with every electromagnetic source other than the true source of life and light and thereby gives glory to these sources grieves God as well as His Holy Spirit. He just wants to love us, and wants us to love Him with the same love, but we, as a race, are in love with ourselves, glorifying each other or wanting the credit ourselves.

"*Because, although they knew God, they did not glorify Him as God, nor were thankful, but became futile in their thoughts, and their foolish hearts were darkened. Professing to be wise, they became fools*" (Romans 1:21-22). But it is through God's grace that He provides us with little gifts, like a fluorescent or phosphorescent object that is in the similitude of His glory, for truly the full manifestation of His glory would destroy us, as it clearly indicates in Exodus 33:20-23: "*But He said, 'You cannot see My face; for no man shall see Me, and live.' And the Lord said, 'Here is a place by Me, and you shall stand on the rock. So it shall be, while My glory passes by, that I will put you in the cleft of the rock, and will cover you with My hand while I pass by. Then I will take away My hand, and you shall see My back; but My face shall not be seen.'*"

Revelation 1:14 communicates the reason: "*His head and hair were white like wool, as white as snow, and His eyes like a flame of fire.*" His eyes are as lasers, with precision-tool burning power. "*For the eyes of the Lord run to and fro throughout the whole earth, to show Himself strong on behalf of those whose heart is loyal to Him*" (2 Chronicles 16:9). His laser-like eyes are the light source to which we may resonate, and the interaction causes a manifestation of His glory.

That the glory of God is patterned after fluorescence and phosphorescence provides the elusive phenomenon of glory a solid foundation. Having dominion over the entire electromagnetic spectrum, the frequency of light emitted by God is unlimited. When we are in a quiet and solid relationship with Him, many times the light frequency that He emits will be of a commensurate value to that within us. At other times, however, depending upon the appointment, the light that He emits will be of a greater frequency, and our subsequent absorption will cause us to jump to a higher energy level and then be subdued by God's peace and fluoresce, exhibiting His glory.

Dependent upon His will as well as our specific appointment, however, our absorption of His light may cause us to enter a state of prolonged absorption which also characterizes His rest, and a corresponding phosphorescent glory may linger for hours: "*When Christ who is our life appears, then you also will appear with Him in glory*" (Colossians 3:4). While our physical body doesn't fluoresce, our spirit does and that light often will shine through the eyes.

Corresponding to the fact that every element and thus every creature have their own spectral emission, there is likewise a specific glory that is characteristic of every life form.

> *All flesh is not the same flesh, but there is one kind of flesh of men, another flesh of animals, another of fish, and another of birds. There are also celestial bodies and terrestrial bodies; but the glory of the celestial is one, and the glory of the terrestrial is another. There is one glory of the sun, another glory of the moon, and another glory of the stars; for one star differs from another star in glory. So also is the resurrection of the dead. The body is sown in corruption, it is raised in incorruption. It is sown in dishonor, it is raised in glory. It is sown in weakness, it is raised in power. It is sown a natural body, it is raised a spiritual body. There is a natural body, and there is a spiritual body. And so it is written, "The first man Adam became a living being." The last Adam became a life-giving spirit (1 Corinthians 15:39-45).*

Due to the different frequencies that correspond with the fruit of the Spirit and its anti-fruit with which each of us can connect, our light emissions will differ as well as fluctuate. Accordingly, the spectral frequency that God emits which corresponds with love will be absorbed and reemitted according to *our own* resonant absorption. And if the light within us is darkness, then rather than absorbing His light and reemitting it, we will only absorb it, which corresponds with the absorption of a black hole. As "*God is love*" (1 John 4:8) and love is as white light, when the fruit of love is in operation there is resonance between God and the one with the fruit, and the vibrational character of the light provides a glow.

Chapter 38
Photochemical Faith

The white light of love, which has been given to all men in that it is equivalent to life as opposed to death, is that which enables man to fulfill his purpose on earth. "*What does the Lord your God require of you, but to fear the Lord your God, to walk in all His ways and to love Him, to serve the Lord your God with all your heart and with all your soul*" (Deuteronomy 10:12). To connect with God's light, which emanates from heaven at a ninety degree angle of righteousness, presupposes that a receptor seed for the Word of God has taken root in the heart. The receptor seed in turn will receive and resonate with all corresponding frequencies, which will bear witness to the record-bearers in heaven—the Father, the Word, and the Holy Ghost. The requirements in Deuteronomy 10:12 address only one thing—full absorption of God's white light that *enables* man to assume God's very nature. Contrary to the process whereby man is forced to conform to other men's ways, by way of the righteousness that comes through faith may undergo a process that parallels the weak interaction, whereby he may become a new creature in Christ Jesus (see 2 Corinthians 5:17).

As reception of the white light of God's love provides man with the opportunity to operate according to faith, which comes by hearing the Word of God and by which a man may be accounted as righteous, a connection may be drawn between the Word of God, love, and faith. Adhering to a process similar to photosynthesis which converts the sun's light into chemical energy[8], faith converts the Word's energy into "*the substance of things hoped for, the evidence of things not seen.*"

Accordingly, we may understand the Word of God in the context of chemical energy which *is* the potential to be transformed itself or to affect changes in other substances. And we know that both conditions apply, as John 1:14 attests that "*the Word became flesh and dwelt among us*" and John 1:3-4 declares that "*all things were made through Him, and without Him nothing was made that was made. In Him was life, and the life was the light of men.*" Deductively paraphrasing, we could then say, "In the beginning all things were made by the Word of God (a presumed spiritual vaporous substance), which was God, and in the Word was life, which was the light of men." Conclusively then, the light of men was

8 "Photochemistry," IUPAC Gold Book, Modified August 19, 2012, accessed January 28, 2014, http://goldbook.iupac.org/P04588.html.

embedded within the spiritual substance of the Word, which through the process of deposition, as in the settling of particles or vapor onto a preexisting surface, transitioned into a solid phase of matter.

Therefore in the beginning the light that men emitted (our inherent spectral emissions) had its origin in the Word of God, which was God. However, with the local magnetic forces vying for man's attention, many have been persuaded to turn from the one true global magnetic force of God and His Word, and by turning from God's magnetic field there is an automatic change in the electric field to which one becomes connected. Herein lies the fundamental foundational exchange between us and the resonant frequencies available and the consequent glory that is able to be imparted.

With consideration that all things were made by the Word of God in which the light of men was embedded, at the same time that the Word became flesh and dwelt among men, the conclusion that the Word of God behaves as both a seed and as the water that waters the seed may be logically drawn. As water, which remains unaffected itself, causes or accelerates a chemical reaction within a seed, behaving as a catalyst in this context, "*the washing of water by the word*" as mentioned in Ephesians 5:26 functions accordingly. Essentially, this *washing of water by the Word* is the precise process that occurred during the dispensations detailed in Genesis 1 and 2, which refer to the former and the latter rain, respectively.

According to God's way of operating by calling those things which are not as though they are (see Romans 4:17), the initial seeds of creation were conceived in faith and were called into being before they actually revealed themselves in physical form. They were then watered by the Word of God when they were *let* loose from their state of bondage to the conceptual realm in Genesis 1; this constituted both the "former rain" as well as their creation. Genesis 2 then constituted the "latter rain" where the initial seeds were watered and brought forth.

What becomes evident is the vital role that faith plays in ensuring that an initial seed will bud, grow, and produce a harvest. And because the Word, like a seed, can be converted into another form, the Word behaves as a chemical substance; similarly, due to the fact that spoken words in general have an effect on the hearts and minds of the hearers, the action of speaking can be considered an initiator of a chemical reaction. Faith, which "*comes by hearing and hearing by the Word of God,*" therefore depends upon the chemical reaction that occurs when the Word is spoken. Therefore, while "*faith is the substance of things hoped for, the evidence of things not seen,*" the Word of God must be heard in order for the chemical reaction of faith to occur: "*So then faith comes by hearing and hearing by the Word of God*" (Romans 10:17). This "*hearing, and hearing by the Word of God"* then corresponds to the watering of the water by the Word, which causes an initial seed to bud.

However, while the Word of God, Jesus, may be understood as functioning like a seed as well as a catalyst, let us not forget that as the *triple point* of God, the Word of God also functions as a mediating phase, as does snow. In addition, the Word enables us to give up our thoughts in exchange for God's thoughts, and thereby provides a true means by which we may become the "*salt of the earth,*" which enables us to enter into a covenant of salt. These examples of the operatives of God simply express how and why God has remained so elusive to so many.

Chapter 39
Silver and Gold

The shimmer and shine of silver and gold appears on the surface gloriously. However, when ultraviolet or gamma rays are absorbed beyond the surface level, these metals exhibit luminescence. As if emulating God's glory, when silver and gold particles are infused in soda-lime silicate glasses and have been exposed to synchrotron irradiation, gold exhibits a bright brilliant green luminescence, and silver, a brilliant white. When embedded in glass, which is often made by melting sand, gold and silver display a pattern of glory. With regard to silver's and gold's figurative worth, in speaking with the church at Corinth, Paul proclaimed:

> *According to the grace of God which was given to me, as a wise master builder I have laid the foundation, and another builds on it. But let each one take heed how he builds on it. For no other foundation can anyone lay than that which is laid, which is Jesus Christ. Now if anyone builds on this foundation with gold, silver, precious stones, wood, hay, straw, each one's work will become clear; for the Day will declare it, because it will be revealed by fire; and the fire will test each one's work, of what sort it is (1 Corinthians 3:10-13).*

It may be interesting to note that glass is often made by melting sand, and in Isaiah 10:22, God's people are referred to as sand: *"For though your people, O Israel, be as the sand of the sea, a remnant of them will return."* If a man's "work" shall be tried with fire, and God's people are referred to as sand, the logical result of the trial by fire would be glass. And in Revelation 15:2 it says, *"And I saw something like a sea of glass mingled with fire, and those who have the victory over the beast, over his image and over his mark and over the number of his name* [666, see Revelation 13:18], *standing on the sea of glass, having harps of God."*

That there will be a division between those who truly love God and those who simply proclaim they do is spoken about in Romans 9:6-9: *"But it is not that the word of God has taken no effect. For they are not all Israel who are of Israel, nor are they all children because they are the seed of Abraham; but, 'In Isaac your seed shall be called.' That is, those who are the children of the flesh, these are not the children of God; but the children of the promise are counted as the seed. For this is the word of promise: 'At this time I will come and Sarah shall have a son.'"* While this definitively speaks about those in the Middle East, it also applies to all who accept Jesus as their Lord and Savior, because through Jesus we may be accounted as Abraham's seed through the avenue of faith, because Isaac was the "seed" born by faith.

In light of this apparent division, the origin of which was between Isaac and Ishmael—the son of the bondwoman, Hagar—those standing on the glass and the glass itself may represent Abraham's seed. Signifying the free (Isaac) and the bound (Ishmael), they who had "*gotten the victory over the beast who stand on the sea of glass"* may in fact signify the seed of Isaac, or those who believe in Jesus by faith. The bound, however, represented by the sea of glass and perhaps representing the seed of Ishmael, are those who resist the Word of truth and *"whose end is to be burned"* (Hebrews 6:8).

With the implicit understanding that anything tried by fire will be burned and the level to which it is destroyed depends upon the material, out of the referenced building materials in 1 Corinthians 3:10-13, silver and gold represent the most enduring substances. For these materials, fire simply signifies a means of purification; for the others, however, it signifies destruction. Gold and silver are strong enough to go through the refiner's fire: *"But who can endure the day of His coming? And who can stand when He appears? For He is like a refiner's fire and like launderers' soap. He will sit as a refiner and a purifier of silver; He will purify the sons of Levi, and purge them as gold and silver, that they may offer to the Lord an offering in righteousness"* (Malachi 3:2-3).

Silver and gold are thus described as desirable building materials, relating not to ornaments for the outer body/building, but rather to the integrity of the inner life and to the heart's intentions/condition. As substances that serve as examples that emulate God's glory, both silver and gold are good electrical and thermal conductors which relates to the ability to impart the Spirit and the Word of God to others. They are also good reflectors even of infrared frequencies, which would imply a quality similar to white light. The capability of reflecting other frequencies is an essential property of white light, as it is in fact the representative power of love. Accordingly, when gold is mentioned in the Bible as the most precious of metals, it is referring to the qualities that emulate God, because "*God is love"* (1 John 4:8).

"Silver and gold I do not have, but what I do have I give you: In the name of Jesus Christ of Nazareth, rise up and walk" (Acts 3:6). This was spoken by Peter when he encountered a crippled man who sat begging at the gate of the temple. *"And [Peter] took him by the right hand and lifted him up, and immediately his feet and ankle bones received strength. So he, leaping up, stood and walked and entered the temple with them—walking, leaping, and praising God. And all the*

people saw him walking and praising God. Then they knew that it was he who sat begging alms at the Beautiful Gate of the temple" (Act 3:7-10).

Then, however, everyone made a mistake that many people make; they began to marvel at the man's healing and began to look at Peter and John as if they themselves had power.

> *So when Peter saw it, he responded to the people: "Men of Israel, why do you marvel at this? Or why look so intently at us, as though by our own power or godliness we had made this man walk? The God of Abraham, Isaac, and Jacob, the God of our fathers, glorified His Servant Jesus, whom you delivered up and denied in the presence of Pilate, when he was determined to let Him go. But you denied the Holy One and the Just, and asked for a murderer to be granted to you, and killed the Prince of life, whom God raised from the dead, of which we are witnesses. And His name, through faith in His name, has made this man strong, whom you see and know. Yes, the faith which comes through Him has given him this perfect soundness in the presence of you all (Acts 3:12-16).*

The precious blood of Jesus, which was shed for all of us, has enabled the Holy Spirit to come and dwell within those who believe that Jesus is the Son of God, which allows them to become sons of God themselves: *"For as many as are led by the Spirit of God, these are sons of God"* (Romans 8:14). We have consequently been enabled to assume a new freeborn, firstborn identity; we are no longer slaves to the law of entropy, which governs the processes of decay, and are considered heirs *"of God through Christ"* (Galatians 4:7). Through Jesus, the Word of God, we are provided the truth about the identity of being a new creature in Christ Jesus (see 2 Corinthians 5:17), and we are enabled through the Spirit to walk in newness of life, and through faith obtain a "righteous" standing:

> *For the law of the Spirit of life in Christ Jesus has made me free from the law of sin and death. For what the law could not do in that it was weak through the flesh, God did by sending His own Son in the likeness of sinful flesh, on account of sin: He condemned sin in the flesh, that the righteous requirement of the law might be fulfilled in us who do not walk according to the flesh but according to the Spirit. For those who live according to the flesh set their minds on the things of the flesh, but those who live according to the Spirit, the things of the Spirit (Romans 8:2-5).*

"Now the works of the flesh are evident, which are: adultery, fornication, uncleanness, lewdness, idolatry, sorcery, hatred, contentions, jealousies, outbursts of wrath, selfish ambitions, dissensions, heresies, envy, murders, drunkenness, revelries, and the like; of which I tell you beforehand, just as I also told you in time past, that those who practice such things will not inherit the kingdom of

God" (Galatians 5:19). The kingdom of God is an inheritance bestowed by a righteous God who, while full of grace, also measures our hearts according to a plumb line and is able to take full account because, *"That which cometh out of the man, that defileth the man. For from within, out of the heart of men, proceed evil thoughts, adulteries, fornications, murders, thefts, covetousness, wickedness, deceit, lasciviousness, an evil eye, blasphemy, pride, foolishness: all these evil things come from within, and defile the man"* (Mark 7:20-23 KJV).

Understanding that the heart of man is the base from which all of these things originate signals the importance of maintaining the ground of one's heart in which receptor seeds are sown. For these receptor seeds connect to either the invisible Holy Spirit of God and to His kingdom or to any number of frequencies in the atmospheric realm, which correspond to the kingdom of this world. Referring to the Holy Spirit, Luke 17:20-21 says, *"The kingdom of God cometh not with observation...for, behold, the kingdom of God is within you"* (KJV). Likewise it may be said that the kingdom of this world comes not with observation for, behold, the kingdom of this world is within you; these kingdoms consequently represent the coupling of receptor seeds and the frequencies to which they connect.

Without the Holy Spirit, we remain bound to the works of the flesh, but with faith in the name of Jesus, the true treasure of God is able to be imparted, which enables us to be free. And through His power, we, like Peter, may impart the *"Word of His power"* (Hebrews 1:3) to those in need of strength. Nonetheless, that gold and silver have been equated to this holy treasure is significant. Sharing the properties of ductility and malleability, these metals are both flexible and enduring and are able to withstand stresses like being drawn into thin wires or being hammered into thin sheets. Jesus Himself certainly exhibited this quality, and there are many other overcomers who have been in commensurate straits.

Additionally, these two metals are relatively unaffected by air, moisture, and most corrosive substances, with gold being the least reactive of metals, holding onto its electrons so as to remain pure. And although gold is among the softest of metals, it has the capacity to outlast other elements during weathering. Perhaps most important though is these metals' tendency to resist corrosion and oxidation, which relates to Matthew 6:20-21: *"But lay up for yourselves treasures in heaven, where neither moth nor rust destroys and where thieves do not break in and steal. For where your treasure is, there your heart will be also."*

In reflection of the grace of God, as silver and gold belong to the group of noble metals, which signal their relative rarity, their precious value parallels that of God's chosen people, Israel:

> *The Lord did not set his love upon you, nor choose you, because ye were more in number than any people; for ye were the fewest of all people:*

but because the Lord loved you, and because he would keep the oath which he had sworn unto your fathers, hath the Lord brought you out with a mighty hand, and redeemed you out of the house of bondmen, from the hand of Pharaoh king of Egypt. Know therefore that the Lord thy God, he is God, the faithful God, which keepeth covenant and mercy with them that love him and keep his commandments to a thousand generations (Deuteronomy 7:7-9 KJV).

The resistance to corrosion parallels a heavenly pattern that God has used to get a message across to those who will hear: As corrosion and oxidation both occur when electrons are donated or lost from an atom, *LotK* would conclude that these processes are equivalent to losing one's "*mind of Christ*" (1 Corinthians 2:16), or in other words allowing one's thoughts to stray from what Philippians 4:8 says: *"Finally, brethren, whatever things are true, whatever things are noble, whatever things are just, whatever things are pure, whatever things are lovely, whatever things are of good report, if there is any virtue and if there is anything praiseworthy—meditate on these things."* And it is with this in mind that we can see why the world is in such a state that it is; our world lingers on bad reports as well as on all of the dishonest, unjust, false, and immoral things within it.

At the appointed time when Jesus returns for those who love Him, they will "*see Him as He is,*" because they too will have sublimed, having been *"raised incorruptible*" like snow, or like gold in its purified state. However, in order to be raised incorruptible, accounted as pure gold, acceptance of the Spirit of God into one's life is essential, for this is the receipt of the purchase that Jesus made when He took His position on the cross: *"For you were bought at a price; therefore glorify God in your body and in your spirit, which are God's*" (1 Corinthians 6:20). This "redemption" speaks about the command not to sacrifice firstborn men and unclean animals, but to pay a price for them instead: "*Nevertheless the firstborn of man you shall surely redeem, and the firstborn of unclean animals you shall redeem. And those redeemed of the devoted things you shall redeem when one month old, according to your valuation, for five shekels of silver, according to the shekel of the sanctuary, which is twenty gerahs*" (Numbers 18:15-16).

While gold and silver shekels were not exchanged for our redemption, Jesus paid for us with His own blood. Sacrificing Himself for our sins as an ultimate sin offering, and like a wave offering and a heave offering as well, He became the spiritual symbol of grace, given to all freely with the offer of enabling people to enter into covenant with Him. Numbers 18 provides an Old Testament/Covenant parallel to Jesus' New Testament/Covenant: Speaking to Aaron, the Lord proclaimed, "*This also is yours: the heave offering of their gift, with all the wave offerings of the children of Israel; I have given them to you, and your sons and daughters with you, as an ordinance forever. Everyone who is clean in your house may eat it"* (Numbers 18:11). Aaron and his sons were priests and as such were in the position of receiving and eating the children of Israel's sacrifices.

Under the new covenant—of which Jesus, the Word of God, stands as a Mediator between man and Father God—however, those who receive the white light of God's love are accounted as priests themselves: "*Jesus Christ, the faithful witness, the firstborn from the dead, and the ruler over the kings of the earth. To Him who loved us and washed us from our sins in His own blood, and* ***has made us kings and priests*** *to His God and Father, to Him be glory and dominion forever and ever"* (Revelation 1:5-6). Jesus Christ, the Word of God, has made us equal to kings and priests, which constitutes a new identity and implies that part of our inheritance includes eating of the sacrifices. Herein lies clarification of any confusion with which Matthew 26:26 might be met: "*And as they were eating, Jesus took bread, blessed and broke it, and gave it to the disciples and said, 'Take, eat; this is My body.'"* By partaking of communion, therefore, we simply are acknowledging our true identity as priests, who have as their inheritance the meat of the sacrifices made to God the Father on behalf of the people.

As Jesus' alternate identity is the Word of God, however, "eating" bread that represented the body of Jesus as Matthew 26:26 implies has a greater implication, as is written in Matthew 4:4: "*Man shall not live by bread alone, but by every word that proceeds from the mouth of God."* Treasured as silver and gold that does not easily corrode, the redemptive power of Jesus and His Word brings us out of *darkness: "But you are a chosen generation, a royal priesthood, a holy nation, His own special people, that you may proclaim the praises of Him who called you out of darkness into His marvelous light"* (1 Peter 2:9); "*having been born again, not of corruptible seed but incorruptible, through the word of God which lives and abides forever*" (1 Peter 1:23).

Chapter 40
Catalytic Faith

As "*God is love*" (1 John 4:8) and love is as white light, when the fruit of love is in operation in the heart, not only is a perpendicular (righteous) connection possible with God, but depending upon our own walk with God and/or our own energy level, the vibrational character of the light can provide a luminescent (inner) glow. Corresponding closely and harmoniously, the resonance that occurs between the Word, faith, and love may most accurately be discerned as a photochemical reaction. As the heavenly pattern of the photosynthetic process, which converts the sun's light into chemical energy, the Word of God, in which is embedded the light of God, is converted into the chemical energy of faith, which is substantiated by Romans 10:17: "*faith comes by hearing, and hearing by the Word of God.*"

In addition, as John 1:4 declares that "*in [the Word] was life; and the life was the light of men,*" light is embedded in the Word of God in the same way that a nucleus is embedded in an atom, and is therefore considered central and essential to the entity's life force. As such, in the process of conversion, the chemical energy of the *developing* faith mixes with the Word, and similar to how light reflects light in a laser cavity, faith in the Word and the Word reflect each other until a critical point is reached. This point can also be seen as the moment of conception, such as when a sperm penetrates an egg and initiates the process of division or fission. This process of conversion, however, most specifically entails a chemical reaction having taken place, which releases the chemical energy of faith; the energy thus may be understood as being released when the Word of God is heard and consequently understood, as Hebrews 4:2 proclaims: "*For unto us was the gospel preached, as well as unto them: but the word preached did not profit them, not being mixed with faith in them that heard it.*"

This, in turn, occurs when receptor seeds for the Word have been planted in the heart, such as when John the Baptist came and preached repentance. John was sent for one reason: He came to "*prepare the way of the Lord; make His paths straight*" (Matthew 3:3) by preparing men's hearts with receptor seeds for the Word of God that would specifically lead them to repentance. By preaching that the time of turning one's heart from old ways and accepting the offer of a new start had arrived, and that there was one who would come after him, John planted receptor seeds that would resonate with the Word of God when they heard it. Consequently, if they had "*ears to hear*," there would be a corresponding chemical reaction and a subsequent connection with God's ninety-degree angle of the white light of love. And through this connection, by which the grace of God was made evident through both the chemical substance and chemical energy of faith, those with "*ears to hear*" would respond and be accounted as righteous. Nonetheless, this process is to be understood as a photochemical reaction with the light of God's love as the primary catalyst.

As light appears to be embedded in the Word of God, "*In Him [the Word] was life; and the life was the light of men*" (John 1:4), we can more fully understand how the process of creation relies on light, which has been substantiated by theorists such as Emmy Noerther. If light is embedded in the Word similar to how neutrons are embedded in a nucleus, speaking the Word would be equivalent to *loosing* the Word from a potentially *stuck* state. This would initiate a fission event with the consequent release of light, which is intimately involved in the creation of matter. John 1:1,3 substantiates this claim: "*In the beginning was the Word, and the Word was with God, and the Word was God. All things were made through Him, and without Him nothing was made that was made.*" This makes sense when we consider that the light of men, which is the life, is embedded in the Word of God; when the Word goes forth, the light goes forth as well.

This presents an illustration of a potential chain reaction that occurred in the very beginning with light acting as the initiating substance: Embedded within the Word of God, light was sent forth as an initiator of creation, which created matter that could undergo fission when it absorbed the Word of God. Exemplifying a cyclical process, this nuclear-like chain reaction not only verifies how the Word and light were used to create as both the Bible and theorists contend, but it also explains why we have an ever-expanding universe. "*And God said* (sent forth acoustic waves into the atmosphere, which were coupled with light as thunder and lightning are similarly coupled), *'Let there be light'; and there was light.*" With consideration that this particular light broke through the darkness, and therefore is confirmed to be *light* light as opposed to *dark* light (see Matthew 6:23), its reflective nature is implicit, and what becomes evident is a continual process of self-reflection and thus expansion.

Due to the fact that the Father, the Word, and the Holy Ghost are one and thereby reflect one another, light's self-reflecting process is consequently a heavenly pattern; in effect, the process is evident in the manifestation of a laser. As with a laser, when light propagates through an optical resonance cavity, the light gains its strength by the mechanism of self-reflection. The same thing occurs as the light of the Word of God (functioning as water) interacts with the Word (functioning as seed) of God within one's heart, and is ushered forth out one's mouth. Light reflects light and resonates back and forth, again and again, until the whole inner cavity of one's heart is filled with light, which, consequently, is embedded in the Word of God; this explains the fruitfulness of meditation in the Word. And herein lies a description of how light and sound are coupled together and are amplified through a process of self-reflection.

As the only species that has been privileged with the capability of self-reflection as well as speech, human beings send forth the coupled waves of sound and light primarily in response to having been created in the image of God. While man can send forth sound waves that become coupled with visible waves, many of these coupled waves actually occur simply as conceptual vibrations within our heart as we meditate upon words, which are heard inaudibly. This was the source of the conceptual seeds that were *birthed* or *loosed* from a state of being bound to the darkness, which was in dominant control in the beginning, as mentioned in Genesis 1. The conceptual vibrations, in turn, are coupled with the fruit of the Spirit or its anti-fruit and accordingly will determine the light frequencies that we emit as well as concurrent feelings of attraction and repulsion.

When the Word of God or words in general are uttered, it indicates that a thought has already been conceived. The vocalization of words then may be considered equivalent to the products of nuclear fission—words are *birthed* in the same way that free neutrons (word carriers), as well as radioactive light emissions, are emitted when a nucleus splits. And when references to seed occur with regard to David's or Abraham's heirs (see John 7:42) or when the Word of God is itself referred to as seed as in Luke 8:11, the comparative association appears to pan out. As lightning precedes thunder, the conception precedes the articulation of what has been conceived. Accordingly, the creatures of the earth were *created* before God had caused a mist to water the whole face of the ground (see Genesis 2:6), which consequently initiated the process of division or fission.

Interestingly enough, it was not until after the ground had been watered and fission had been initiated that what had been created was actually formed and provided with names: "*And out of the ground the Lord God formed every beast of the field, and every fowl of the air; and brought them unto Adam to see what he would call them: and whatsoever Adam called every living creature, that was the name thereof*" (Genesis 2:19 KJV). While God calls those things that are not as though they were (see Romans 4:17), meaning that He operated by faith, Adam operated primarily by sight. And perhaps this is a primary distinguishing factor between the first Adam who was bound to the flesh and the second Adam, Jesus, who was free and one with the Spirit, as First Corinthians suggests:

> *There is a natural body, and there is a spiritual body. And so it is written, "The first man Adam became a living being." The last Adam became a life-giving spirit. The first man was of the earth, made of dust; the second Man is the Lord from heaven. As was the man of dust, so also*

> *are those who are made of dust; and as is the heavenly Man, so also are those who are heavenly. Now this I say, brethren, that flesh and blood cannot inherit the kingdom of God; nor does corruption inherit incorruption. Behold, I tell you a mystery: We shall not all sleep, but we shall all be changed—in a moment, in the twinkling of an eye (1 Corinthians 15:44-45, 47-48, 50-52).*

With regard to the process of fission many things apply, but its fundamental involvement in the creative force is of particular interest due to God's identity as Creator. As creativity involves the absorption of new thoughts that travel on frequencies that instigate more new thoughts, we can assume that this frequency corresponds with white, reflective light as opposed to dark, absorbing light. What is required, therefore, is an initiation as well as a propagation of new thoughts. However, in order for the initiation of creativity to occur in the first place, a new thought must be absorbed into the system, like a neutron is absorbed by a uranium nucleus with the end result being the birth of a new idea. Revelation 21:5 mentions, "*Then He who sat on the throne said, 'Behold, I make all things new.'*"

In reference to God as this maker of all things new, or in other words, the Creator, what becomes solidly evident is that God is ultimately open-minded, understanding of every difference, and non-judgmental. He is openly accepting of new ideas and new perspectives, which are in truth not new to Him at all, as He is, "*Alpha and the Omega, the Beginning and the End, the First and the Last*" (Revelation 22:13). "*Declaring the end from the beginning, and from ancient times the things that are not yet done, saying, My counsel shall stand, and I will do all my pleasure*" (Isaiah 46:10 KJV). God knows and understands everything and can find out all, as Proverb 8:12 substantiates, *"I wisdom dwell with prudence, and find out knowledge of witty inventions"* (KJV).

In acknowledgement of the fact that thoughts precede words and words precede action, in order for new things to be *made* in the physical realm, they need to have been *created* in the invisible realm first, or in other words they need to have been conceived. Similar to the moment when a sperm penetrates an egg, when a new thought is absorbed into a formerly closed system, the response of division, which parallels fission, looses thoughts from their bound state. Therefore, unless there is openness to new ways of perceiving and interpreting the world, which comes by *understanding* others' differences and not judging or fearing them, thoughts, like seeds, are self-similar and will therefore be readily replicated. Adhering to Isaac Newton's laws of motion, without openness to new ways of being and perceiving, which dislodges cyclically driven thoughts, these will remain traveling in their closed-loop cycles indefinitely. Unless a new thought is accepted and absorbed into the system, therefore, similar to the way that an electron is captured during beta-plus decay, the identity of the thoughts will not change and there will be no subsequent creativity. Similarly, in the same way that a uranium nucleus absorbs a free neutron and undergoes fission, the absorption of a new thought/conception instigates a splitting of a closed-minded box, so to speak. And in this way a nuclear reactor may be seen as an incubator of new thought processes.

Nonetheless, as new thoughts are ushered into a system and ultimately cause fission to occur, paralleling the free neutrons as well as the radioactive light that are produced in the natural, the products of *new thought* absorption are words. And whether audible or not, words uttered even silently cause vibration, which may be understood as simply a response to a previously-absorbed *light* frequency. Now, while sound and visible light both exist on the continuum of the electromagnetic spectrum, this particular light frequency that is coupled with sound may be regarded as abiding in the realm referred to in Genesis 1:4 when God divided the light from darkness, calling the light Day and calling the darkness Night. While this light reflects the love of God, this darkness may most readily be regarded as referring to the absorbing property of fear and hatred. And in addition to how the emitting of light corresponds with the absorption of acoustic vibrations, the absorption of words effectively stimulates the release of words, and the consequent release of explosive power, which may correspond with either faith or fear.

In this regard, faith may be considered the actual chemical substance in a reaction, which like gold is malleable and ductile in molecular composition and can reflect the golden glory of God's Word and transform accordingly. Like gold, too, it is enduring and able to withstand the fire; like the light, it is multiplied until a laser-like power can be emitted, which when coupled with the Word of God has explosive creative power.

As with all transitions from the invisible to the visible, the chemical substance of faith (invisible) requires a domain of resistant energy to be in operation. This is where works come in—Noah, Abraham, and Moses all resisted their sense knowledge that would cause them to doubt God's Word. For example, *"By faith Abraham obeyed when he was called to go out to the place which he would receive as an inheritance. And he went out, not knowing where he was going"* (Hebrews 11:8). While most of us would shudder at the potential judgment of such an act, Abraham simply obeyed out of confident faith in Him whom He had heard. Sense knowledge or reasonability stands with defiant resistance against such obedience, and that's the problem. The *working* chemical substance of faith requires more confidence in God's Word than what is placed in the world's wisdom and ways.

According to this reasoning, it becomes evident that for faith (the chemical substance) to transition into the substance of things hoped for and become the evidence of things not seen (see Hebrews 11:1), the Word of God must be heard, listened to, and absorbed, with the ultimate effect of causing a *loosing* of words and power. *"So then faith comes by hearing, and hearing by the Word of God"* (Romans 10:17), which when truly heard will be accompanied by a reemitting of a similar light frequency, as well as words proclaiming faithfulness, such as Romans 10:8-10 indicates: *"'The word is near you, in your mouth and in your heart' (that is, the word of faith which we preach): that if you confess with your mouth the Lord Jesus and believe in your heart that God has raised Him from the dead, you will be saved. For with the heart one believes unto righteousness, and with the mouth confession is made unto salvation."*

As the process that causes light to chemically alter matter, photosynthesis thereby relates to the pattern whereby the light of God's Word interacts with the substance of faith. Like a seed hidden in the ground bursts with radioactive power after a period of heat absorption, the substance of faith sprouts with new life after a period of absorbing the Word of God. Faith, the substance of things hoped for, comes by hearing and hearing by the Word of God, and the manifestation of faith as well as the sprouted seed is equated to the manifestation of the glory of God. John 12:23 references this equation: *"The hour is come, that the Son of man should be glorified,"* followed by the example in verse 24, "*Except a corn of wheat fall into the ground [be planted] and die, it abideth alone: but if it die, it bringeth forth much fruit*" (KJV). This suggests that the glory is manifested only after the potential within a seed bursts forth.

When the glorification of Jesus is mentioned, therefore, it is referring to the time when the manifestation of His full potential, or what He was born to accomplish, had been satisfied. The potential that awaits the children of God is the rest that God desires His people to enter: "*There remains therefore a rest for the people of God. For he who has entered His rest has himself also ceased from his works as God did from His*" (Hebrews 4:9-10). That gold and silver manifest luminescence when exposed to light of a particular frequency relates to the treasure of potential that God desires His people to actualize by building on the foundation of salvation by faith in Jesus Christ with enduring and fruitful substance.

While the Word of God, Jesus, functions as a catalyst, let us not forget that as a hypothetical triple point of God, the Word of God also holds treasures like that of snow, a mediating force that is uniquely manifested. In addition, regarding salt, which is formed by the union of two oppositely charged ions, the Word of God functions as a negatively charged ion (anion) that seeks to bind with us, who in turn represent positively charged cations. The result of this union, therefore, is that we assume the identity of the "*salt of the earth*," bound with Him as in a covenant of salt. These examples of the operatives of God simply express how and why God has remained so elusive to so many. As governor of all the forces, His roles, offices, and domains operate according to circular causality, where everything is connected and contingent on everything else, a fundamental universal feedback loop.

"[God] has in these last days spoken to us by His Son, whom He has appointed heir of all things, through whom also He made the worlds; who being the brightness of His glory and the express image of His person, and upholding all things by the word of His power" (Hebrews 1:2-3). That the kingdom of God is upheld by the *"Word of his power"* implies that God's power is as a force with the Word operating as a boson, or force carrier, just as a photon is a carrier of the electromagnetic force. With this *LotK* proposes a fifth force: Alongside electromagnetism, the strong force, the weak force, and gravity, is the force of faith, with a corresponding and mediating force carrier: The Word of God.

Chapter 41
LotK Omega

As the mediating force carrier, the Word of God—whose primary alternate identities are God, Jesus the Messiah, a seed, and the Sword of the Spirit—in one sense behaves as does a photon, a gluon, a W boson and the proposed graviton, the hypothetical elementary particle that mediates the force of gravitation. As a catalyst that initiates and carries out substantial transformations, along with light, the Word of God carries the capacity to transform the molecular composition of any substrate, or material in or upon which it is deposited. As a seed, the Word of God maintains a potential energy commensurate to the power released in a nuclear fission event; as the sword of the Spirit, its binding and loosing power is equivalent to that of a laser. As alluded to in Matthew 16:19, the Word of God, as the mediating *particle* of the force of faith, was given to us as it was to Peter, representative of the "*keys of the Kingdom of heaven.*" Jesus asked Simon Peter:

> *"But who do you say that I am?" Simon Peter answered and said, "You are the Christ, the Son of the living God." Jesus answered and said to him, "Blessed are you, Simon Bar-Jonah, for flesh and blood has not revealed this to you, but My Father who is in heaven. And I also say to you that you are Peter, and on this rock I will build My church, and the gates of Hades shall not prevail against it. And I will give you the keys of the kingdom of heaven, and whatever you bind on earth will be bound in heaven, and whatever you loose on earth will be loosed in heaven" (Matthew 16:16-19).*

The keys thus signified the binding and loosing power of the Word of God, which binds up the broken-hearted (see Isaiah 61:1) as well as "*binds the strong man*" from entering into one's house (Matthew 12:29) and loosens the prisoners from their bonds, as well as the power of strongholds (see Isaiah 42:7 and 2 Corinthians 10:4). However, the reason why Peter was given these keys in the first place was because he had received knowledge from God, and had believed by faith that which he had received was true; he had heard, received, and believed, and he had not received it from men, but directly from the Holy Spirit. This represents "the rock" upon which Jesus has built His church—a Spirit-filled body.

There's so much more, but three topics in particular stand out at the moment as having not been addressed in this book: The first, the Urim and Thummim that are mentioned in Exodus 28:28-30, Leviticus 8:8, Deuteronomy 33:8, and 1 Samuel 23:9. These were objects that were to be bound by rings to the ephod with a blue lace on the breastplate of judgment worn by Aaron when he went before the Lord. Interestingly, these two items sound very much like uranium and thorium, two of the most radioactive elements known to man, which decay primarily through alpha decay or through the emitting of helium atoms. Appearing to serve as symbols of the Lord's power, these items are mentioned in Deuteronomy 33:8: "*Let Your Thummim and Your Urim be with Your holy one, whom You tested at Massah.*" In this Scripture these items appear to be emblems of protection, and in First Samuel again they seem to serve as protecting aids. And when David needed protection from Saul who wanted to kill him, "*he said to Abiathar the priest, 'Bring the ephod here'*" (1 Samuel 23:9) on which these items were adhered.

Perhaps most striking of all, however, is 1 Samuel 28:6, which says: "*And when Saul inquired of the Lord, the Lord did not answer him, either by dreams or by Urim or by the prophets,*" which seems to indicate that Urim (uranium, a radioactive element) is a manifestation of the power of God. In addition, thorium serves as a radioactive shield as does depleted uranium. It appears that perhaps the Urim and the Thummim together serve perhaps more as weapons, whereas the Thummim may have protected Aaron's heart from any radioactive fallout inherent in God when he went into the Holy of Holies, before the presence of the Lord. This also would validate the assertion that God Himself is fundamentally electromagnetic in nature.

A second subject is related to the Word of God being equated to food. According to Matthew 4:4: "*It is written, 'Man shall not live by bread alone, but by every word that proceeds from the mouth of God.'*" Hereby, food and the Word appear related. The interesting aspect of this is that the metabolizing of *real* food parallels the metabolizing of the Word of God. As food stores heat energy which is liberated when it is digested, so too does the Word of God. Every food can be evaluated according to its efficiency. Food efficiency, in turn, is a measure of how efficiently a food is converted to energy. Foods with a high efficiency are easily converted into energy while low efficiency foods are not.

The applicability of this is that when anyone hears the Word of God and receives it into their heart, they gain energy because that source of fuel has lingered in the mind and heart and has been converted into fuel. Conversely, when the Word is heard but not truly received by the heart, one could equate that to food that has been metabolized too quickly, in other words, "in one ear and out the other." And when the Word is not heard, it doesn't undergo any metabolism at all, and similar to food that is not metabolized, the weight just piles on. This appears to be related to Hebrews 12:1-2: "*Therefore we also, since we are surrounded by so great a cloud of witnesses, let us lay aside every weight, and the sin which so easily ensnares us, and let us run with endurance the race that is set before us, looking unto Jesus, the author and finisher of our faith, who for the joy that was set before Him endured the cross, despising the shame, and has sat down at the right hand of the throne of God.*" Someone without ears to hear, therefore, will not be capable of laying aside every weight; neither will they be able to run the race.

There's so much more; *LotK* has only scratched the surface, and as 1 Corinthians 13:9-10 says, "*For we know in part and we prophesy in part. But when that which is perfect has come, then that which is in part will be done away.*"

I thought as a child.
In the polite aftermath
Of idle chatter,
Adults would question,
"What do you do?"
"I thought—as a child
But...I [have] put away childish things."
(1 Corinthians 13:11)

Note from the Author

Logics of the Kingdom is *not* a theory for which I can take any credit; it was rather a commissioned assignment. God knew that my mission in life is to be on board with the plan that He has for my life as well as for His kingdom, and that I would be obedient to His call: *Logics of the Kingdom* was therefore "co-missioned."

"*He who has ears to hear, let him hear*": if you will simply call out to Him as a child, and by the way, in the deepest crevices of all of our hearts, we are still the same three-year-old child that we were so many years ago. We all just want to be loved, accepted, told that we're okay, respected, and understood. Truly, you'll never know anyone who will embrace you and love you unconditionally the way the Lord Jesus Christ does and will. Romans 10:9 proclaims the simplicity that is in Christ Jesus: "*If you will confess with your mouth the Lord Jesus* [proclaim that He is who He says He is] *and believe in your heart that God has raised Him from the dead, you will be saved,"* or in other words accepted by Him as one of His own. *"Nevertheless the foundation of God standeth sure, having this seal, The Lord knoweth them that are his. And, let every one that nameth the name of Christ depart from iniquity* (2 Timothy 2:19 KJV).

With the deepest respect,

H.B. Fortinberry

LotK Summary Points

- All things in heaven and on earth are established according to patterns which have been established from the inception of the world.
- Every scientific discovery made by man provides evidence of God's operational means by which people may uncover the truths about the governing operations in heaven.
- Light creates, and it is this creative power which is embedded in the Word of God.
- The Word of God is Jesus; therefore, the light of men, or the life force is embedded in Jesus, is in His name and in His Word.
- God governs all of the fundamental forces, but may be understood as operating primarily as the electromagnetic force.
- People are subject to the electrical frequencies that operate on the vast continuum of the electromagnetic spectrum; the principalities, powers, rulers of darkness of this world, and the spiritual wickedness in high places (see Ephesians 6:12) lie on one end of this continuum, while God's primary frequency lies on the other.
- As *"God is love"* (1 John 4:8), the frequency that carries His power may be understood as the white light of love, which lies perpendicular to the magnetic field.
- We have been provided the capability of consciously resisting both the fruit of the Spirit as well as its anti-fruit, with the resistance of the fruit of the Spirit being easier due to the natural flow of entropy and time.
- Man's gift of free will and his subsequent ability to choose enables him to make one fundamental choice: Whom he will serve—either life or death, blessing or cursing.
- According to this one choice, all other connections with frequencies will ultimately be made according to *receptor seeds* that pick up specific signals; the analogy of doors or channels, therefore, is most accurate.

> *"I call heaven and earth to record this day against you, that I have set before you life and death, blessing and cursing: therefore choose life, that both thou and thy seed may live" (Deuteronomy 30:19 KJV).*

Index

About the Author

Heather Fortinberry is a bridge builder, drawing connections between scientific, psychological, and social inquiry and the Word of God, as revealed in scripture. At just seventeen years of age she had been in a coma. Since that experience, Heather spent years studying scientific processes and the fundamental forces of the universe.

Using the Bible as her source of strength combined with her extensive knowledge, Heather describes physical phenomena as manifestations of spiritual principles and proves how the fundamental theory lays a foundation upon which understanding can be built. In 2014 she published this book, *Logics of the Kingdom*, and in 2020 she released *Logics of the Kingdom Dynamics*.

Heather is a life coach whose primary motivation is to provide people with a sound explanation of the world around them with the hope that knowledge and understanding will be gained and that they might fulfill their potential. Heather has a master's degree in psychology and taught for ten years at Brook's Institute in Ventura, California. For more information, please contact hfortinberrylifecoaching.com or logicsofthekingdom.com.

Made in the USA
Monee, IL
04 November 2026

67383843R00144